REVISION CARDS FOR THE NEBOSH NATIONAL GENERAL CERTIFICATE

REVISION CARDS FOR THE NEBOSH NATIONAL GENERAL CERTIFICATE

Dr Ed Ferrett, C Eng, CMIOSH

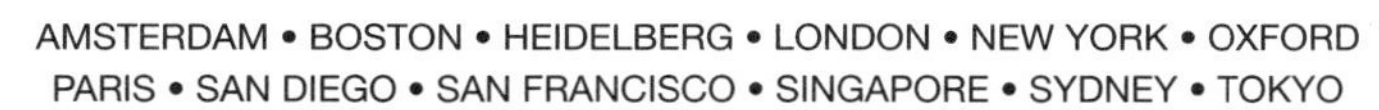

Butterworth-Heinemann is an imprint of Elsevier

Butterworth-Heinemann is an imprint of Elsevier
Linacre House, Jordan Hill, Oxford OX2 8DP, UK
30 Corporate Drive, Suite 400, Burlington, MA 01803, USA

First edition 2009

British Library Cataloguing in Publication Data
A catalogue record for this book is available from the British Library

Library of Congress Cataloging in Publication Data
A catalog record for this book is available from the Library of Congress

ISBN: 978-1-85617-702-3

For information on all Butterworth-Heinemann publications visit our website at www.elsevierdirect.com

Printed and bound in Italy

09 10 10 9 8 7 6 5 4 3 2 1

Table of contents

Preface

Welcome to the NEBOSH National General Certificate Cards from Elsevier/Butterworth-Heinemann. The cards have been designed to be used together with the NEBOSH National General Certificate syllabus and the textbook 'Introduction to Health and Safety at Work' by Hughes and Ferrett. The cards have the following features:

- Revision notes for each of the 8 elements of the two units – NGC1 Management of health and safety and NGC2 Controlling workplace hazards.
- A summary of the learning outcomes and key points is given for each element.
- Important diagrams are included to help revision.

The compact size of the cards ensures that they can be easily carried and used for revision at any time or place. They should be used throughout the course alongside the textbook and course handouts.

Good luck with your NEBOSH course.

Unit NGC1
Management of Health and Safety

Element 1
Foundations of Health and Safety

Learning outcomes

- Outline the scope and nature of occupational health and safety
- Explain briefly the moral, legal and financial reasons for promoting good standards of health and safety
- Outline the legal framework for the regulation of health and safety
- Describe the roles and powers of enforcement agencies, the judiciary and external agencies
- Identify the nature and key sources of health and safety information
- Outline the key elements of a health and safety management system

Key revision points

- The definitions of hazard, risk, civil law, criminal law, common law and statute law
- The employer's duty of care and common law and statutory duties
- Criminal liabilities and defences
- Civil liabilities and defences (particularly negligence)
- The business case for health and safety (direct, indirect, insured and uninsured costs)
- The legal framework – the Health and Safety at Work Act, the Management of Health and Safety at Work Regulations, absolute and qualified duties
- Sources of health and safety information
- Key elements of the health and safety management system HSG 65

Definitions

Accident – an unplanned event that results in damage, loss or harm	**Other definitions:**
Hazard – the potential of something causing harm	**Welfare** – provision of facilities to maintain health and well-being of people in the workplace (e.g. washing, sanitary and first-aid facilities)
Risk – the likelihood of something to cause harm	**Residual risk** – remaining risks after controls applied
Civil law – duties of individuals to each other	**Near miss and dangerous occurrence**
Criminal law – duties of individuals to the State	
Common law – law based on court judgements	
Statute law – law based on Acts of Parliament	

Health: The promotion of normal well being in what they do within their role/work.

Safety: Provision of controls & protection to ensure the good health & well being of personnel.

Dangerous Occurrance: a 'near miss' which could have resulted in severe injury/fatality. Reportable under RIDDOR.

Drivers for good health and safety management

Moral drivers	Legal drivers	Financial drivers
Ethical reasons to reduce: ➤ accident rates ➤ industrial disease and ill-health rates	**Poor management can lead to:** ➤ prosecutions ➤ civil actions – compensation claims	**Poor health and safety management can lead to:** ➤ direct and ➤ indirect costs

employer liability insurance:
ELI.

must have to cover cost of incident
no matter what employers financial
state is.

Legal framework

Sub-divisions of law	
Criminal law	**Civil law**
➤ enforced by the State to punish individuals (and/or organizations) ➤ individual is prosecuted by an Agency of the State (e.g. HSE, Local Authorities, Fire Authority) ➤ individual(s) guilty or not guilty ➤ courts can fine or imprisonment ➤ proof 'beyond reasonable doubt' ➤ cannot insure against punishment	➤ deals with disputes between individuals (and/or organizations) to address a civil wrong (tort) ➤ individual(s) and/or organizations are sued ➤ individual(s) are liable or not liable ➤ courts can award compensation and costs ➤ proof is based on 'balance of probabilities' ➤ employers must insure against civil actions (*Employers Liability Insurance*)
Sources of law	
Common law	**Statute law**
➤ based on judgements made by courts (strictly judges in courts) ➤ generally courts bound by earlier judgements (precedents) ➤ lower courts follow judgements of higher courts ➤ in health and safety definitions of negligence, duties of care and terms such as practicable and reasonably practicable are based on common law judgements	➤ law laid down by Acts of Parliament ➤ Health and Safety at Work Act 1974 ➤ specific duties mainly in Regulations or Statutory Instruments ➤ takes precedence over common law

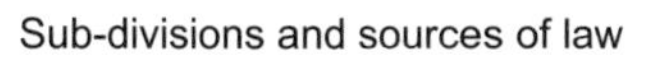

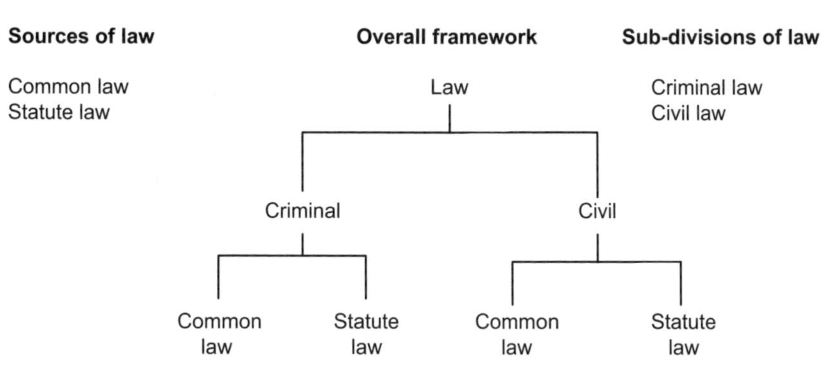

Figure 1 The relationship between the sub-divisions and sources of law

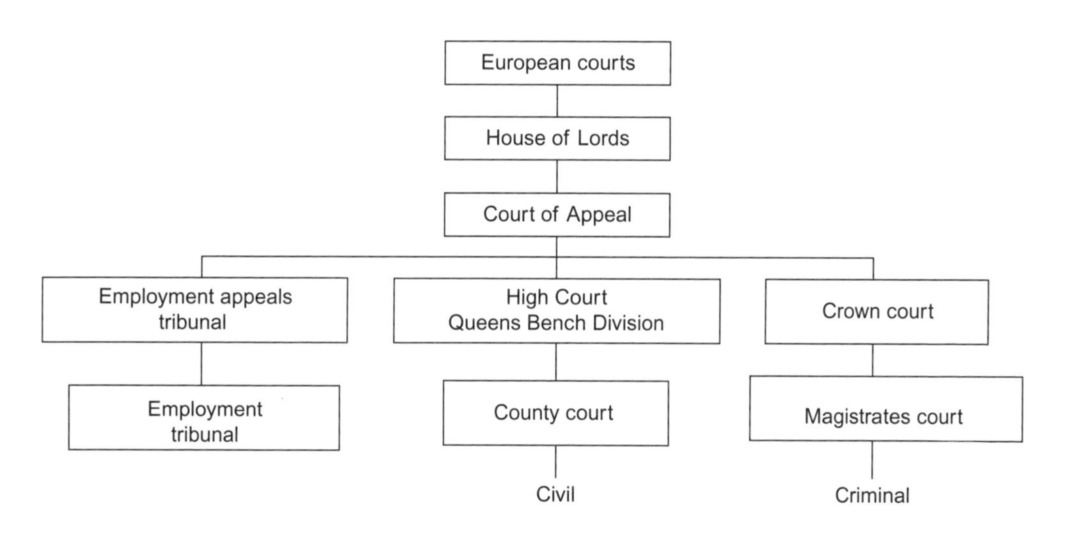

Figure 2 The Court system for health and safety in England and Wales

Aspects of civil law

Employer's common law duty of care

The employer has a duty of care to each of his employees. This duty cannot be assigned to another person. The duty of care falls into four groups and the employer must provide:

- a safe place of work, including access and egress
- safe plant and equipment
- a safe system of work
- safe and competent fellow employees and adequate levels of supervision, information, instruction and training

Employees have a right not to be harmed in any way by their work and they are expected to take reasonable care in their workplace.

Negligence

- lack of reasonable care or conduct resulting in injury, damage or loss
- must be reasonably foreseeable that the acts or omissions could result in injury

injury is direct cause of breach.

Vicarious liability

– when the defendant is acting in the normal course of his employment during the alleged incident, the defence of the action is transferred to his employer

Levels of statutory duty

ABSOLUTE – The employer MUST comply with the law. Regulations use the verbs 'must' or 'shall'

PRACTICABLE – The employer must comply if it is technically possible. Difficulty, inconvenience or cost cannot be taken into account

REASONABLY PRACTICABLE – If the risk is small compared with the cost, time and effort required to further reduce the risk, then no action is required

Defences against negligence claims	Partial defences
➤ a duty of care was not owed ➤ there was no breach of the duty of care ➤ any breach of a duty of care did not result in the specific injury, disease, damage and/or loss suffered	➤ contributory negligence – employee contributed to the negligent act ➤ volenti no fit injuria – the risk was willingly accepted by the employee ➤ Acts of God – riot, terrorist event, etc.

The legal framework for the regulation of health and safety

Influence of the European Union (EU)

EU role is to harmonize health and safety standards across all member states

Powers of EU in health and safety mainly derived from the Treaty of Rome (1957) and the Single European Act (1986)

Article 95A (was 100A) – health and safety standards for plant and equipment

Article 138A (was 118A) – minimum health and safety standards in employment

European Directives set out the specific minimum aims of a given area of health and safety

European Directives must be incorporated into the national law of all member states

First introduction into UK law was in 1993 with the 'European Six Pack'

European Court of Justice rules on interpretation of EU law

European Court of Human Rights interprets European Human Rights Convention

Health and Safety at Work Act 1974

Background topics

The Health and Safety at Work Act (HSW Act) was introduced in 1974 and is a criminal law based on the recommendations of the Robens Report

Main recommendations of the Robens Report

1. A single Act to cover all workers containing general duties
2. The Act should cover everyone affected by the employer's undertaking
3. Emphasis on management of health and safety, including training and supervision
4. Encouragement of employee involvement in accident prevention
5. Enforcement should be targeted at 'self-regulation' rather than prosecution

HSW Act is an *Enabling Act* allowing regulations to be produced by the HSE for the Secretary of State without further Acts of Parliament being required

HSW Act contains mainly general duties with specific duties defined in Regulations

Regulations

- state the law
- often implement EU Directives
- breaches are criminal offences possibly leading to enforcement action
- describe the minimum health and safety standards that need to be achieved
- usually apply across all organizations but sometimes apply to specific industries

Approved Codes of Practice (ACOPs)

- supplementary practical interpretation of regulations that give more detail on the regulatory requirements
- special legal status – quasi-legal because it may be possible to comply with regulations by some other more effective practice
- COPs are legally binding if the particular regulations indicate that they are quoted in an Enforcement Notice

Guidance

- two forms of guidance – legal and best practice both available as HSE publications
- not legally binding
- gives more informative and practical advice than COPs
- also often available as British Standards and as industrial or trade guidance

Role and activities of the HSE

- advise on the development of regulations
- enforce health and safety regulations
- provide information to organizations (ACOPs, guidance notes, leaflets and other publications, accident and ill-health statistics)
- launch national health and safety campaigns on specific topics
- carry out accident and other investigations
- offer advice to employers and others on statutory duties
- take enforcement action
- instigate criminal proceedings and publicize organizations that receive enforcement notices

The Health and Safety at Work Act

Section 2 – Duties of employers to employees

General duty – ensure, as far as is reasonably practicable, the health, safety and welfare of all employees

Specific duties:

- ➤ safe plant and systems of work
- ➤ safe use, handling, transport and storage of substances and articles
- ➤ provision of information, instruction, training and supervision
- ➤ safe place of work, access and egress
- ➤ safe working environment with adequate welfare facilities
- ➤ a written safety policy together with organizational and other arrangements (if 5 or more employees)
- ➤ consultation with safety representatives and formation of safety committees where there are recognized trade unions

Section 3 – Duties of employers to others affected by their undertaking

– 'others' could include contractors , general public, visitors, patients, students

Section 4 – Duties of landlords or owners

- safe access and egress for those using the premises
- buildings and any equipment supplied with them is safe

Section 6 – Duties of suppliers

Suppliers (including designers) of articles and substances for use at work to ensure, as far as is reasonably practicable, that

- articles are designed and constructed to be safe and without risk to health at all times when they are being set, cleaned, used and maintained
- substances are similarly safe and without risk to health when being used, handled, stored or transported
- arrange, where necessary, for suitable testing and examination
- supply suitable safety information and any revisions to customers

Section 7 – Duties of employees

- take care for the health and safety of themselves and others who may be affected by their acts or omissions
- co-operate with their employer, as far as is necessary, to ensure compliance with any statutory health and safety duty

Section 8

No person to misuse anything provided for health, safety or welfare purposes

Section 9

Employees cannot be charged for health and safety requirements

Enforcement of the Act

Powers of inspectors

- enter premises at any reasonable time, accompanied by a police officer, if necessary
- examine, investigate and require the premises to be left undisturbed
- access to all records and other relevant documents
- take samples, photographs and, if necessary, dismantle and remove equipment or substances
- seize, destroy or render harmless any substance or article
- take statements
- issue improvement and prohibition notices and, possibly, a formal caution
- initiate prosecutions

Fire & Rescue authority
Ammendment notice.

Enforcement notices

Improvement notice

Issued for a specific breach of the law

Appeal within 21 days to an Employment Tribunal – notice suspended until appeal is heard or withdrawn

Prohibition notice

Issued to halt an activity that could lead to serious injury

Appeal within 21 days to an Employment Tribunal – notice is not suspended

Deferred prohibition notice – stops the work activity within a specified time limit

Penalties

Summary offences	Indictable offences
➤ for most health and safety offences up to £20,000 fine and/or up to 6 months imprisonment	➤ unlimited fines for all health and safety offences
➤ in the future, the maximum period of imprisonment will be increased to 12 months ➤ up to 5 years disqualification for convicted directors	➤ up to 2 years imprisonment for all health and safety offences ➤ up to 15 years disqualification for convicted directors

Management of Health and Safety at Work Regulations

Employers' duties

- undertake suitable and sufficient written risk assessments (5 or more employees)
- put in place effective health and safety management arrangements and record them if 5 or more employees
- employ a competent health and safety person
- develop suitable emergency arrangements and inform employees and others
- provide health and safety information to employees and others, such as other employers, the self-employed and their employees who are sharing the same workplace and parents of child employees or those on work experience
- co-operate in health and safety matters with other employers who share the same workplace
- provide employees with adequate and relevant health and safety training
- provide temporary workers and their contract agency with appropriate health and safety information
- protect new and expectant mothers and young persons from particular risks
- under certain circumstances, as outlined in Regulation 6, provide health surveillance for employees

The information that should be supplied by employers under the Regulations is:

- risks identified by any risk assessments including those notified to him by other employers sharing the same workplace
- the preventative and protective measures that are in place
- the emergency arrangements and procedures and the names of those responsible for the implementation of the procedures

Employees' duties

- ➤ use any equipment or substance in accordance with any training or instruction given by the employer
- ➤ report to the employer any serious or imminent danger
- ➤ report any shortcomings in the employer's protective health and safety arrangements

Other external agencies

- ➤ Fire and Rescue Authority
  - enforce fire safety law
  - undertake random fire inspections (often to examine fire risk assessments)
  - can issue alteration, improvement and prohibition notices
  - need to be informed during the planning stage of building alterations when fire safety of the building may be affected

- Environment Agency (Scottish Environmental Agency)
  - responsible for authorizing and regulating emissions from industry
  - ensuring effective controls of the most polluting industries
  - monitoring radioactive releases from nuclear sites
  - ensuring that discharges to controlled waters are at acceptable levels
  - setting standards and issuing permits for the collection, transporting, processing and disposal of waste (including radioactive waste)
  - enforcement of the Producer Responsibility Obligations (Packaging Waste) Regulations 1997
  - enforcement of the Waste Electrical and Electronic Equipment (WEEE) Directive and its associated directives

- Insurance Companies
  - legal requirement for employers to insure against liability for injury or disease to their employees arising out of their employment
  - also offer fire and public liability insurance
  - can influence health and safety standards by weighing the premium offered to an organization against its health and safety record

Sources of information on health and safety

Internal sources

- accident and ill-health records and investigation reports
- absentee records
- inspection and audit reports undertaken by the organization and by external organizations such as the HSE
- maintenance, risk assessment (including COSHH) and training records
- documents which provide information to workers
- any equipment examination or test reports

External sources

- Health and Safety legislation
- HSE publications, such as ACOPs, guidance documents, leaflets, journals, books and their website
- International (e.g. ILO), European and British standards
- health and safety magazines and journals
- information published by trade associations, employer organizations and trade unions
- specialist technical and legal publications
- information and data from manufacturers and suppliers
- the Internet and encyclopaedias

Costs of accidents and ill health

Direct costs

– directly related to the accident and may be insured or uninsured

Insured direct costs normally include:

- claims on employers and public liability insurance
- damage to buildings, equipment or vehicles
- any attributable production and/or general business loss

Uninsured direct costs include:

- fines resulting from prosecution by the enforcement authority
- sick pay
- some damage to product, equipment, vehicles or process not directly attributable to the accident (e.g. caused by replacement staff)
- increases in insurance premiums resulting from the accident
- any compensation not covered by the insurance policy due to an excess agreed between the employer and the insurance company
- legal representation following any compensation claim

Indirect costs

– costs which may not be directly attributable to the accident but may result from a series of accidents. Again these may be insured or uninsured

Insured indirect costs include:

- a cumulative business loss
- product or process liability claims
- recruitment of replacement staff

Uninsured indirect costs include:

- loss of goodwill and a poor corporate image
- accident investigation time and any subsequent remedial action required
- production delays
- extra overtime payments
- lost time for other employees, such as a First Aider, who attend to the needs of the injured person
- the recruitment and training of replacement staff
- additional administration time incurred
- first aid provision and training
- lower employee morale possibly leading to reduced productivity

Some of these items, such as business loss, may be uninsurable or too prohibitively expensive to insure. Therefore, insurance policies can never cover all of the costs of an accident or disease either because some items are not covered by the policy or the insurance excess is greater than the particular item costs.

Framework for Health and Safety Management HSG 65

The six elements of HSG 65 are:

1. Policy – A clear health and safety policy contributes to business efficiency and continuous improvement throughout the operation. The policy should state the intentions of the organization in terms of clear aims, objectives, targets and senior management involvement
2. Organizing – A well-defined health and safety organization should identify health and safety responsibilities at all levels of the organization. An effective organization will be noted for good communication, the promotion of competency, the commitment of all employees and a responsive reporting system
3. Planning and implementing – A clear health and safety plan based on risk assessment, sets and implements performance standards, targets and procedures through an effective health and safety management system. The plan should set priorities and objectives for the control or elimination of hazards and the reduction of risks
4. Measuring performance – This includes both active (sometimes called proactive) and reactive monitoring of the health and safety management system. It is also important to measure the organization against its own long-term goals and objectives
5. Reviewing performance – The results of monitoring and independent audits should indicate whether the objectives and targets set in the health and safety policy need to be changed. Changes in the health and safety environment in the organization, such as an accident, should also trigger a performance review. Performance reviews should include comparisons with internal performance indicators and the external performance indicators of similar organizations with exemplary practices and high standards
6. Auditing – An independent and structured audit of all parts of the health and safety management system reinforces the review process. If the audit is to be really effective, it must assess both the compliance with stated procedures and the performance in the workplace. It will identify weaknesses in the health and safety policy and procedures and identify unrealistic or inadequate standards and targets

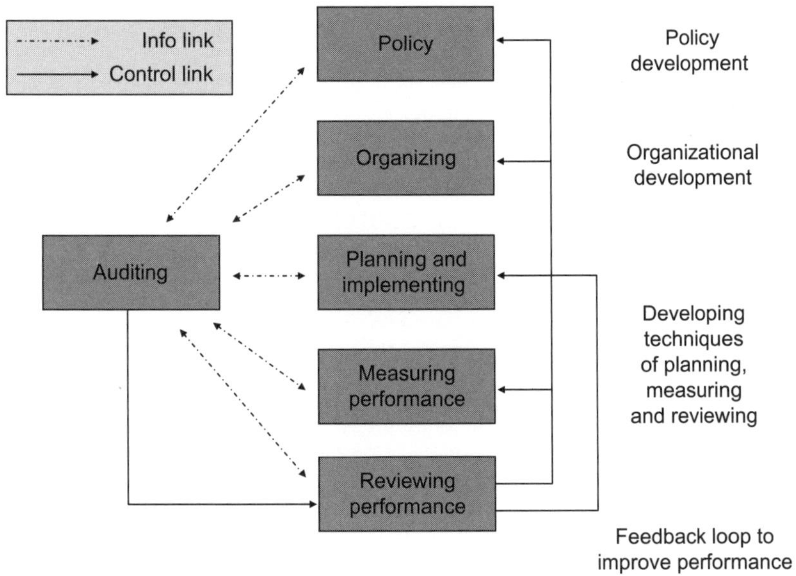

Figure 3 The framework for Health and Safety Management HSG 65

Element 2
Policy

Learning outcomes

- Explain the purpose and importance of setting policy for health and safety
- Describe the key features and appropriate content of an effective health and safety policy

Key revision points

- The legal requirements for a health and safety policy
- Key elements of a health and safety policy
- Target setting for health and safety performance
- Circumstances leading to the need for a review of health and safety policy

Legal requirements

- Section 2(3) of HSW Act requires a written health and safety policy when there are 5 or more employees
- a policy statement of intent giving aims and measurable objectives of the organization
- details of health and safety organization including individual duties and responsibilities
- details of health and safety arrangements in terms of systems and procedures

Policy statement of intent

- includes aims and objectives for health, safety and welfare
- includes duties of employer and employees to each other and the wider public and others
- includes performance targets for the immediate and long-term future
- includes the name and post of the person responsible for the management of health and safety in the organization
- statement should be signed and dated by the most senior person in the organization
- should include a date by which the statement will be reviewed
- statement should be posted on prominent notice boards throughout the workplace

Organization includes	Arrangements include
➤ organizational chart ➤ identification of main hazards ➤ responsibilities and duties ➤ safety monitoring and audit system and procedures ➤ health and safety training and information ➤ accident investigation teams ➤ allocation of resources for health and safety including finance ➤ health surveillance organization ➤ terms of reference for the safety committee ➤ compilation of safety manual	➤ employee health and safety code of practice ➤ main risk assessments ➤ maintenance procedures ➤ fire precautions ➤ emergency procedures ➤ first aid arrangements ➤ accident reporting procedures ➤ exposure control to hazardous substances and other physical hazards ➤ procedures for machinery safety including safe systems of work ➤ electrical safety procedures ➤ catering and food hygiene procedures ➤ contractor and visitor procedures ➤ procedures for consultation with employees ➤ procedures for waste disposal

Reasons for a review of the health and safety policy

- significant organizational changes have taken place
- there have been changes in personnel and/or legislation
- health and safety performance has fallen below the occupational group's benchmarks
- the monitoring of risk assessments and/or accident/incident investigations indicate that the health and safety policy is no longer totally effective
- enforcement action has been taken by the HSE or Local Authority
- a sufficient period of time has elapsed since the previous review

The effects of a positive health and safety performance are to:

- support the overall development of personnel
- improve communication and consultation throughout the organization
- minimize financial losses due to accidents and ill-health and other incidents
- directly involve senior managers in all levels of the organization
- improve supervision, particularly for young persons and those on occupational training courses
- improve production processes
- improve the public image of the organization or company

The reasons for unsuccessful health and safety policies include:

- the statements in the policy and the health and safety priorities not understood by or properly communicated to the workforce
- minimal resources made available for the implementation of the policy
- too much emphasis on rules for employees and too little on management policy
- a lack of parity with other activities of the organization (such as finance)
- lack of senior management involvement in health and safety
- inadequate personal protective equipment
- unsafe and poorly maintained machinery and equipment
- a lack of health and safety monitoring procedures

Element 3
Organizing for Health and Safety

Learning outcomes

- Outline the legal and organizational health and safety roles and responsibilities of employers, managers, supervisors, employees and other relevant parties
- Explain the requirements placed on employers to consult with their employees

Key revision points

- The health and safety responsibilities and duties of employers to their employees and others affected by their undertaking
- The health and safety responsibilities of directors, managers and supervisors
- The competence and responsibilities of the health and safety practitioner/adviser
- The duties and responsibilities between client and contractor to ensure a high standard of health and safety during the contract
- The duties of employers to consult with the workforce

Employer responsibilities

Under the HSW Act, the employer has a duty to safeguard the health and safety and welfare at work of:

- employees
- other workers – agency, temporary or casual
- trainees
- contractors
- visitors
- neighbours and the general public

Key actions required of the employer are to:

- ensure competent advice on health and safety matters
- obtain current Employers' Liability insurance and display the certificate
- compile a health and safety policy and ensure that an adequate health and safety management system is in place
- ensure that risk assessments of all work activities are undertaken and any required controls are put in place
- provide the workforce with health and safety information and training
- provide adequate welfare facilities
- consult workforce on health and safety issues
- report and investigate some accidents, diseases and dangerous occurrences
- display prominently the health and safety law poster (or supply appropriate leaflet)

Other employer responsibilities

Visitors and the general public

Possible hazards

- unfamiliarity with the workplace processes
- lack of knowledge of site layout
- unfamiliarity with emergency procedures
- inappropriate personal protective equipment
- inadequate or unsigned walkways
- added vulnerability if young or disabled

Possible controls

- visitor identification (use of badges)
- routine signing in and out
- escorted by a member of staff
- provision of information on hazards and emergency procedures
- given explicit site rules (wearing of personal protective equipment)
- clear marking of walkways

For night workers, employers should:

- determine the normal working time for night workers
- if the working time is more than 8 hours per day on average, determine whether the number of hours can be reduced
- develop an appropriate health assessment and method of making health checks
- ensure that proper records of night workers are maintained, including details of health assessments
- ensure that night workers are not involved in work which is particularly hazardous

Employee responsibilities

These are covered in Element 1

Directors' responsibilities

Directors and board members should ensure that:

- health and safety arrangements are properly resourced
- competent health and safety advice is obtained
- regular reports are received on health and safety performance
- any new or amended health and safety legislation is implemented
- risk assessments are undertaken
- there are regular audits of health and safety management systems and risk control measures
- there is adequate consultation with employees on health and safety issues

Managing Directors/Chief Executives, Line managers and Supervisors play key roles in ensuring that the health and safety policy is delivered and monitored

Health and safety adviser must:

- be competent following the attainment of a health and safety qualification and training
- report directly to a senior management on matters of policy
- keep up to date with technological advances and legislative changes
- advise on the establishment of health and safety, maintenance and accident investigation procedures
- provide liaison with external agencies, such as the HSE, Fire Authorities, contractors, insurance companies and the public

Contractors are protected by the HSW Act (Section 3). The following points should be considered when contractors are employed:

- Health and safety must be included in contract specification
- All significant hazards must be included in specification
- The contractor must be selected with safety in mind
- Prior to the start of work, health and safety policies should be exchanged
- The contractor must be given basic site and health and safety information, such as on welfare and first-aid arrangements, significant hazards, safe storage of chemicals and the name of the contract supervisor
- Where appropriate, the contractor should supply risk assessments and method statements
- For construction work, the contractor should be aware of the position of buried/overhead services and the arrangements for the disposal of waste
- The contractor should be monitored during the progress of the contract by the contract supervisor
- The contract supervisor should check that the work has been completed safely at the end of the contract

Consultation with employees

Safety representatives from recognized trade union have the following rights:

- to investigate accidents and dangerous occurrences
- to investigate health and safety complaints
- to undertake workplace inspections
- to receive information from health and safety inspectors
- to attend health and safety committee meetings
- to have access to suitable facilities to perform their functions
- to be allowed time off with pay for health and safety training

If two or more representatives request in writing for a health and safety committee to be set up, then the employer must comply within 3 months

Representatives of Employee Safety (ROES) – were established under Consultation with Employees Regulations
The employer must *inform* and *consult* employees on health and safety matters

ROES have the following functions:

- inform the employer of health and safety concerns of the workforce
- inform the employer of potential hazards and dangerous occurrences in the workplace
- inform the employer of any general matters that affect the health and safety of the workforce
- speak on behalf of the workforce to health and safety inspectors

Employer must consult on:

- new processes or equipment or any changes in them
- the appointment arrangements for health and safety competent person
- the results of any risk assessments
- the arrangements for the management of health and safety training
- the introduction of new technologies

Types of information that employer does NOT need to disclose if it:

- violates of a legal prohibition
- endangers national security
- relates to a specific individual
- could harm the company commercially
- was obtained from legal proceedings

Element 4
Promoting a Positive Health and Safety Culture

Learning outcomes

- Describe the concept of health and safety culture and its significance in the management of health and safety in an organization
- Identify indicators which could be used to assess the effectiveness of an organization's health and safety culture and recognize factors that could cause its deterioration
- Identify the factors which influence safety related behaviour at work
- Identify methods which could be used to improve the health and safety culture of an organization
- Outline the internal and external influences on an organization's health and safety standards

Key revision points

- The definition and importance of a health and safety culture
- The relationship between culture and performance
- The definition of human factors and their influence on the culture
- The development of a positive health and safety culture
- The importance of good communication
- The different forms of health and safety training
- The internal and external influences on the health and safety culture of an organization

Features of a good health and safety culture

Leadership and commitment to health and safety at all levels
Acceptance that high standards are achievable
Mutual trust throughout the organization
Detailed risk assessments and control and monitoring procedures
Health and safety policy including a code of practice and required health and safety standards
Training, communication and consultation systems
Encouragement to the workforce to report potential hazards
Health and safety monitoring systems
Prompt accident investigation and implementation of remedial actions

Indicators of a health and safety culture

Accident/incident rates
Sickness and absentee rates
Resources available for health and safety
Level of legal and other compliance
Turnover rates for employees
Level of complaints
Selection and management of contractors
Levels and effectiveness of communication and supervision
Health and safety management structure
Level of insurance premiums

Human factors

1 in 10 near misses = 1 accident
90 percent of accidents are due to human error – 70 per cent are due to poor management

Organization, job and personal factors

Organization

- must have a positive health and safety culture
- manage health and safety by providing leadership and involvement of senior managers
- motivate the workforce to improve health and safety performance
- measure health and safety performance

Job

- recognize possibility of human error
- good ergonomics, equipment design and layout of workstation
- clear job description
- safe system of work and operating procedures
- job rotation and regular breaks
- provision of correct tools
- effective training schedule and good communication

Personal factors	Other related factors
The three common psychological factors are: ➤ Attitude – tendency to behave in a particular way in a given situation, influenced by social background and peer pressure ➤ Motivation – the driving force behind the way a person acts or is stimulated to act ➤ Perception – the way in which a person believes or understands information supplied or a situation	➤ self-interest – e.g. effect of bonus systems ➤ position in the team ➤ acknowledgement by management of good work/initiatives ➤ hearing and/or memory loss ➤ experience and competence ➤ age, personality, attitude, language problems ➤ training undertaken and information given ➤ effect of shift working – e.g. night working ➤ health (physical and mental)

Human errors may be:

1. Slips – failure to carry out the correct actions of a task
2. Lapses – failure to carry out particular actions that form part of a working procedure
3. Mistakes are:
   - rule-based – a rule or procedure is applied or remembered incorrectly or
   - knowledge-based – well tried methods or calculation rules are applied incorrectly

Violations may be:

1. Routine – the breaking of a safety rule or procedure is the normal way of working
2. Situational – job pressures at a particular time make rule compliance difficult
3. Exceptional – a safety rule is broken to perform a new task

Development of a positive health and safety culture:

1. Commitment of management is the most important factor:
   - proactive management
   - promotion by example (e.g. wearing personal protective equipment)

2. The promotion of health and safety standards for:
   - selection and design of premises
   - selection and design of plant, processes and substances
   - recruitment of employees and contractors
   - risk assessments and control implementation
   - competence, maintenance and supervision
   - emergency planning and training
   - transportation of the product and its subsequent maintenance and servicing

3. Competence of the workforce including in health and safety
   - knowledge and understanding of the work/job
   - capacity to apply knowledge to the particular task
   - awareness of one's limitations

Communication

- verbal (by mouth) – conversations, telephone
- written – memos, emails, meeting minutes, data sheets
- graphic – safety signs, posters, charts

Barriers to effective communication

- language and dialect
- acronyms and jargon
- various physical and mental disabilities
- attitudes and perception of workers and supervisors

Types of accident propaganda

- statistics
- films, DVDs and posters
- targets
- records

For safety propaganda to be effective, it must have:

- a simple understandable message
- a positive believable message
- an appealing format that will motivate the reader

Types of health and safety training

- Induction – at recruitment
- Job-specific
- Supervisory and management
- Specialist (e.g. first aid)
- Refresher or reinforcement

Internal influences on health and safety culture

- Management commitment
- Production/service demands
- Communication
- Competence
- Employee representation

External influences on health and safety culture

- Expectations of society
- Legislation and enforcement
- Insurance companies
- Trade unions
- State of the economy
- Commercial stakeholders

Element 5
Risk Assessment

Learning outcomes

- Explain the aims and objectives of risk assessment
- Identify hazards by means of workplace inspection and analysis of tasks
- Explain the principles and practice of risk assessment

Key revision points

- The legal requirements for a health and safety risk assessment
- The meaning of 'suitable and sufficient'
- The forms and objectives of risk assessment
- Details of accident categories and types of health risks
- The risk assessment process and its management
- Groups that require a special risk assessment

Legal requirement

Regulation 3 of Management of Health and Safety at Work Regulations requires written suitable and sufficient risk assessment if there are 5 or more employees
Suitable and sufficient means:

- identify significant risks only
- identify measures required to comply with legislation
- remain appropriate and valid over a reasonable period of time

Hazard – the potential to cause harm
Risk – the likelihood to cause harm
Residual risk – the risk remaining after some controls are in place

Forms of risk assessment:	Health risks:
➤ Quantitative – calculated from risk = likeli-hood × severity ➤ Qualitative – descriptor (high, medium or low) used to describe timetable for remedial action ➤ Generic – covers similar activities or work equipment	➤ Chemical – exhaust fumes, paint solvents, asbestos ➤ Biological – legionella, pathogens, hepatitis ➤ Physical – noise vibration, radiation ➤ Psychological – stress, violence ➤ (Ergonomic – musculoskeletal disorders)

Risk assessment process

- Hazard identification – Step 1 of HSE's 5 steps
- Persons at risk – Step 2 of HSE's 5 steps
  - employees, agency/temporar y workers, contractors, shift workers
  - members of the public – visitors, customers, patients, students, children, elderly
  - special groups – young persons, expectant or nursing mothers, workers with a disability, lone workers
- Evaluation of risk level (residual risk) – Step 3 of HSE's 5 steps
  - high, medium and low (defined qualitatively or quantitatively)
  - both occupational and organizational risk levels need to be considered
- Detail risk controls (existing and additional) – Step 3 of HSE's 5 steps
  - the prioritization of risk control is important
  - risks can be reduced at the design stage by using the principles of prevention (see Element 6)
  - risks can be controlled by using the hierarchy of risk control (see Element 6)
- Record of risk assessment findings – Step 4 of HSE's 5 steps
- Monitor and review – Step 5 of HSE's 5 steps – regular reviews required but need to be more frequent if:
  - new legislation introduced
  - new information available on substances or process
  - changes to the workforce – introduction of trainees
  - an accident has occurred

Risk assessment team requirements

- All need training in risk assessment
- Leader should have health and safety experience
- All need to be competent to assess risks in area under examination
- All need to know their own limitations
- Include local line manager in team
- At least one team member with report writing skills

Special cases

Young persons

- are under 18 years
- covered by Regulation 19 MHSWR
- subject to peer pressure and are inexperienced
- are eager to please
- appropriate level and approach of subject matter in training sessions

There are additional requirements if under school leaving age

A special risk assessment must be made to include details of:

- the work activity
- any prohibited processes or equipment
- the health and safety training provided
- the supervision arrangements

Expectant and nursing mothers

There are restrictions on the type of work that can be undertaken

Risks include:

- manual handling
- chemical and biological agents
- ionisng radiation
- passive smoking
- lack of rest room facilities
- temperature variations
- prolonged standing or sitting
- stress and violence to staff

Workers with a disability

- emergency arrangements including 'raising the alarm'
- adequate wheelchair access to fire exit

Lone workers

- special risk assessment (including violence)
- must be fit to work alone
- special training should be given
- possible to handle all equipment and substances alone
- periodic visits by supervisor
- regular mobile phone contact with base
- first aid arrangements
- emergency arrangements

Element 6
Principles of Control

Learning outcomes

- Describe the general principles of control and a basic hierarchy of risk reduction measures that encompass technical, behavioural and procedural controls
- Describe what factors should be considered when developing and implementing a safe system of work for general work activities and explain the key elements of a safe system applied to the particular situations of working in confined spaces and lone working
- Explain the role and function of a permit-to-work system
- Explain the need for emergency procedures and the arrangements for contacting emergency services
- Describe the requirements for, and effective provision of, first aid in the workplace

Key revision points

- The principles of prevention and the hierarchy of risk control
- The control of health risks
- The development and application of safe systems of work and permits-to-work for various activities including those of lone workers
- The development of emergency and first-aid procedures in the workplace

Principles of control

The planning and implementing section of the health and safety management system is based on risk assessment and concerns all the actions taken to control or eliminate hazards and reduce risks. The principles of prevention are used when equipment or processes are being designed or selected. The hierarchy of risk control enables risk to be further controlled.

The principles of prevention are to:	**Hierarchy of risk control:**
➤ avoid risks ➤ evaluate risks which cannot be avoided ➤ adapt work to the individual ➤ adapt to technical changes ➤ replace dangerous items with less dangerous items ➤ develop an overall prevention policy ➤ give priority to collective measures (Safe Place strategy) ➤ give instructions to employees (Safe Person strategy)	➤ elimination ➤ substitution ➤ changing work methods/patterns ➤ reduced time exposure ➤ engineering controls (isolation, insulation and ventilation) ➤ good housekeeping ➤ safe systems of work ➤ training and information ➤ personal protective equipment ➤ welfare ➤ monitoring and supervision ➤ review

Safety signs

Red – prohibition – round – (e.g. no smoking)
Yellow – warning – triangular – (e.g. wet floor)
Blue – mandatory – round – (e.g. ear defenders must be worn)
Green – safe condition-square or rectangular – (e.g. first aid)

Control hierarchy for health risks from hazardous substances

- Change the process or task
- Substitute for a safer substance
- Use substance in a safer form
- Engineering controls – totally enclose
- Engineering controls – use partial enclosure and local exhaust ventilation
- Engineering controls – use general ventilation (dilute ventilation)
- Use safe systems of work and procedures
- Reduce the number of people exposed
- Reduce time exposure
- Use good housekeeping
- Give training and information
- The last resort – use personal protective equipment
- Other controls – use welfare and health surveillance

Method statements are formal written safe systems of work and are often used in construction work

Development of safe systems of work:	Safe systems of work are particularly important for:
1. Assess the task (complexity, accident records, etc.) 2. Identify the significant hazards associated with the task 3. Define safe methods for performing the task (including emergency procedures) – document the methods if required 4. Implement the safe system of work (written safe system to be signed off) 5. Monitor the safe system of work and review it if necessary 6. Train the workforce in safe procedures and enter on training record	➤ maintenance work ➤ contractors ➤ emergency procedures ➤ lone working ➤ vehicle operations ➤ cleaning operations

Permits-to-work are:	Typical responsible persons are:
– *formal* safe systems of work – required to be signed on/off by a responsible person – often require equipment to be locked on/off by a responsible person **to be used whenever there is a high risk of serious injury, such as:** ➤ spaces ➤ live electricity (particularly high voltage) ➤ hot working (e.g. welding work) ➤ some machinery confined maintenance work	➤ site manager ➤ senior authorized person – often the chief engineer ➤ authorized person – issue permits ➤ competent persons – receives permit ➤ operatives – supervised by a competent person ➤ specialists – (e.g. electrical engineer) ➤ engineers – usually responsible for the work ➤ contractors

Confined spaces

Examples include underground chamber, silo, trench, sewer, tunnel

Hazards are:

- lack of oxygen and asphyxiation
- poor ventilation
- presence of fumes
- poor means of access and escape
- drowning
- claustrophobia
- electrical equipment (needs to be flameproof)
- presence of dust (e.g. silos)
- heat and high temperatures
- fire and/or explosion
- poor or artificial lighting

Controls include:

- permit-to-work
- risk assessment
- training and information for all workers entering the confined space
- emergency arrangements in place
- emergency training
- no entry for unauthorized persons

Machinery maintenance work

Hazards are:

- no perceived risk
- no safe system of work
- poor communications
- failure to brief contractors
- lack of familiarity
- poor design

Control of hazards requires:

- effective planning
- a written safe system of work/permit-to-work
- a risk assessment to assess, control and reduce risks
- monitoring to ensure that the system of work and controls are used
- an effective training programme for all involved in the work

Emergency procedures

Examples of types of emergencies:

- fire
- explosions; bomb scares
- escape of toxic gases
- major accident

Typical elements of emergency procedures:

- fire notices and fire procedures (including testing)
- fire drills and evacuation procedures
- assembly and roll call
- arrangements for contacting emergency and rescue services
- provision of information for emergency services
- internal emergency organization – including control of spillages and clean-up arrangements
- media and publicity arrangements
- business continuity arrangements

First aid

Main functions of first-aid treatment:

- preservation of life and/or minimization of the consequences of serious injury until medical help is available
- treatment of minor injuries not needing medical attention

Main first-aid requirements:

- qualified first aiders
- adequate facilities and equipment to administer first aid
- an assessment of required first aid cover and requirements
- an appointed person available to assist first aiders

First-aid box – contents depend on particular workplace needs but should be checked regularly

Basic first aid provision (including number of first aiders) depends on:

- number of workers
- the hazards and risks in the workplace
- accident record and types of injuries
- proximity to emergency medical services
- working patterns (shift work)

***First aider* requires first-aid training:**

- 4-day first-aid course initially
- 2-day refresher course every 3 years

Appointed person has some other first-aid experience/qualification

Element 7
Monitoring, Review and Audit

Learning outcomes

- Outline and differentiate between active (proactive) monitoring procedures, including inspections, sampling, tours and reactive monitoring procedures, explaining their role within a monitoring regime
- Carry out a workplace inspection, and communicate findings in the form of an effective and persuasive report
- Explain the purpose of regular reviews of health and safety performance, the means by which reviews might be undertaken and the criteria that will influence the frequency of such reviews
- Explain the meaning of the term 'health and safety audit' and describe the preparations that may be needed prior to an audit and the information that may be needed during an audit

Key revision points

- The reasons for measuring and monitoring health and safety performance
- The role of standards in the monitoring process
- The differences between reactive and active (proactive) monitoring
- The importance of clear unambiguous report writing
- The role of performance review and audit

Proactive (or active) monitoring (taking action before problems occur) involves:	**Reactive monitoring (taking action after a problem occurs) involves:**
➤ the active monitoring of the workplace for unsafe conditions ➤ the direct observation of workers for unsafe acts ➤ meeting with management and workers to discover any problems ➤ checking documents, such as maintenance records, near miss reports, insurance reports ➤ undertaking workplace inspections, sampling, surveys, tours and audits	➤ review of accident and ill-health reports – often to check that remedial advice has been actioned or ascertain trends and hot spots ➤ review of procedures following dangerous occurrences, other property damage and near misses ➤ review of compensation claims ➤ review of complaints from the workforce and members of the public ➤ review of procedures following enforcement reports and notices ➤ review of risk assessments following the discovery of additional hazards

Workplace inspections – detailed check, often using a checklist, of the whole workplace and should cover:

- the premises (e.g. fire precautions, access/egress, housekeeping)
- the plant, equipment and substances (e.g. machine guarding, tools, ventilation)
- the procedures in place (e.g. safe systems of work, risk assessments, use of personal protective equipment)
- the workforce (e.g. training, information, supervision, health surveillance)

Other issues with inspections:

- the competence of the observers
- the frequency of inspections
- the response to the inspection report
- the use of objective inspection standards

Safety sampling – checking for safety defects in a selected area of the workplace (e.g. all fire extinguishers)
Safety surveys – a detailed inspection of a particular workplace activity throughout an organization (e.g. manual handling)
Safety tours – an unscheduled brief inspection of a work area in the workplace by a team led by a senior manager
Auditing – the independent collection of information on the efficiency, effectiveness and reliability of the whole health and safety management system measured against specific standards. It will check that the following are in place:

- appropriate management arrangements
- adequate risk control systems
- appropriate workplace precautions
- appropriate documentation and records

External audits:	Internal audits:
➤ independence from organization may be questioned ➤ competent ➤ familiar with external benchmarks ➤ more expensive than internal audits ➤ do not know the organization ➤ require more information than internal audits ➤ can offer bland reports	➤ not independent of organization and partiality may be questioned ➤ usually require audit training ➤ less expensive than external audits ➤ know the organization well, particularly critical areas ➤ can spread good practice around the organization ➤ may be unaware of external benchmarks

Audits should take place at regular intervals
Other issues with audits:

Performance review is the final stage of the management process, and reviews:

- all monitoring, inspection and audit reports
- the adequacy of the health and safety management system and performance standards against external benchmarks
- whether new legislation or guidance has been applied
- whether the health and safety policy objectives have been met or need modification to ensure continuous improvement
- whether there has been adequate feedback to/from managers

Element 8

Incident and Accident Investigation, Recording and Reporting

Learning outcomes

- Explain the purpose of, and procedures for, investigating incidents (accidents, cases of work-related ill-health and other occurrences)
- Describe the legal and organizational requirements for recording and reporting such incidents

Key revision points

- The reasons and legal requirements for recording and reporting incidents and accidents
- Basic accident investigation procedures and the different types of accident and incident
- Immediate and underlying causes of incidents
- Issues concerning insurance and compensation claims

Purpose of accident/incident investigation

- to eliminate the cause and future occurrences
- to determine the direct and indirect causes of the accident/incident
- to define any corrective and/or preventative actions
- to identify any deficiencies in risk controls, the health and safety management system and/or procedures
- to ensure that all legal requirements are being met
- to comply with the recommendations of the Wolfe report so that essential information is available in the event of a civil claim

Benefits of accident/incident investigation

- prevention of a recurrence
- prevention of future business losses
- prevention of future increased insurance premiums and costs of criminal and civil actions
- improve employee morale and organization reputation

Causes of accidents

Direct or immediate

- unsafe acts by individuals due to poor behaviour or a lack of training, supervision, information or competence or failure to wear personal protective equipment
- unsafe conditions, such as inadequate guarding, inadequate procedures, hazardous substances, ergonomic and/or environmental factors
- fire or explosive hazards

Indirect, root or underlying

- poor machine maintenance and/or start-up procedures
- management and social pressures
- financial restrictions
- lack of management commitment to health and safety
- poor or lack of health and safety policy and standards
- poor workplace health and safety culture
- workplace and trade customs and attitudes

Elements of an investigation

- Interview relevant operatives, managers, supervisors
- Obtain detailed plans and/or photographs of scene
- Check all relevant records (working procedures, maintenance, training, risk assessments)
- Interview all witnesses
- Possibly arrange for equipment and/or substances to be independently tested
- Produce a concise report for management that includes:
  - background to the accident
  - possible causes of the accident (direct and indirect)
  - relevant health and safety legislation, guidance and standards
  - recommendations (including any remedial actions)
  - any additional training or follow up requirements
- Undertake a post-accident risk assessment

Accident reporting requirements

- the organizational accident/incident report form
- accident book
- RIDDOR 95 – records kept for 3 years

RIDDOR 95

The Regulations apply to employees, self-employed, trainee and contractors. Any fatality or injury to a member of the public resulting in off-site medical treatment is reportable

Fatality

- reportable by quickest means (telephone)
- submit form 2508 within 10 days
- reportable if death occurs within 1 year of accident

Major accident

- reportable as soon as possible
- submit form 2508 within 10 days
- examples – any fracture, other than fingers, thumbs or toes; any amputation: any other injury requiring admittance to hospital for more than 24 hours

Over 3-day lost time injury

- submit form 2508 within 10 days

Disease

- forthwith
- submit form 2508A within 10 days
- examples – hepatitis; occupational dermatitis; hand–arm vibration syndrome

Dangerous occurrence

– a 'near miss' that could lead to serious injury or loss of life
  - reportable by quickest means (telephone)
  - submit form 2508 within 10 days
  - examples – collapse of scaffolding; the collapse, overturning or failure of any load-bearing part of lifts and lifting equipment; explosion or fire resulting in the suspension of normal working for more than 24 hours

Compensation and insurance issues

- importance of documentation and records
- need for pre- and post-accident risk assessments
- impact of Woolf report and documentation required to defend a civil claim

Unit NGC2
Controlling Workplace Hazards

Element 9
Movement of People and Vehicles – Hazards and Control

Learning outcomes

- Identify the hazards that may cause injuries to pedestrians in the workplace and the control measures to reduce the risk of such injuries
- Identify the hazards presented by the movement of vehicles in the workplace and the control measures to reduce the risks they present

Key revision points

- The hazards and control strategies for pedestrian safety
- The hazards and control strategy for safe vehicular operations
- The management of vehicle movements

Hazards to pedestrians include:

- slips, trips and falls on the same level
- falls from height
- collisions with moving vehicles
- being struck by moving, falling or flying objects
- striking against fixed or stationary objects

Slip hazards are caused by:

- wet or dusty floors
- the spillage of wet or dry substances – oil, water, plastic pellets
- loose mats on slippery floors
- wet and/or icy weather conditions
- unsuitable footwear, floor coverings or sloping floors

Trip hazards are caused by:

- loose floorboards or carpets
- obstructions, low walls, low fixtures on the floor
- cables or trailing leads to portable electrical hand tools and other electrical appliances across walkways
- raised telephone and electrical sockets
- rugs and mats – particularly when worn or placed on a polished surface
- poor housekeeping – obstacles left on walkways, rubbish not removed regularly
- poor lighting levels – particularly near steps or other changes in level
- sloping or uneven floors – particularly where there is poor lighting or no handrails
- unsuitable footwear – shoes with a slippery sole or lack of ankle support

Control strategies for pedestrians:

- risk assessment to identify suitable controls
- slip resistant surfaces and reflective edges to stairs and kerbs
- spillage control and drainage
- designated walkways
- fencing and guarding, particularly on stairways
- use of warning signs
- sound storage racking that is inspected and maintained regularly
- maintenance of a safe workplace
- cleaning and good housekeeping procedures
- access and egress
- environmental considerations (heating, lighting, noise and dust)
- special provision for disabled people
- high visibility clothing, appropriate footwear, personal protective equipment (safety harnesses)
- information, instruction, training and supervision

Hazards in vehicle operations include:

- collisions between pedestrians and vehicles
- vehicles crushing feet of pedestrians
- people falling from vehicles
- people being struck by objects falling from or attached to vehicles
- people being struck by or ejected from an overturning vehicle
- communication problems between vehicle drivers and employees or members of the public

Hazards in vehicle operations may be caused by:

- poor working practices, such as the lack of regular vehicular safety checks
- defective maintenance, steering, brakes, tyres and hydraulic hoses
- poor road surfaces and/or poorly drained road surfaces
- overloading of vehicles
- use of unsuitable vehicles to transport people
- inadequate supervision
- poor training or lack of refresher training

Other more general hazards involving pedestrians and vehicles include:

- reversing of vehicles especially inside buildings
- roadways too narrow with insufficient safe parking areas
- roadways poorly marked out and inappropriate or unfamiliar signs used
- too few pedestrian crossing points
- the non-separation of pedestrians and vehicles
- lack of barriers along roadways
- lack of directional and other signs
- poor visibility due to load impeding view, blind corners or inadequate vehicular lighting or mirrors
- poor environmental factors, such as lighting, dust and noise
- ill-defined speed limits and/or speed limits which are not enforced
- vehicles used by untrained and/or unauthorized personnel

Control strategies for safe vehicle operations:

- risk assessment
- designated marked traffic routes with good visibility and well-designed loading and storage areas
- enforced speed limits and/or the fitting of speed governors
- audible warning of approach signals, particularly for reversing
- separation of pedestrians and vehicles
- use of one way systems
- the issue of suitable high visibility clothing and other personal protective equipment to drivers and pedestrians in vehicle operating areas
- separate site access gates for pedestrians and vehicles
- use of pedestrian crossings on vehicular routes
- mirrors on blind corners
- identification of recognized parking and non-parking areas
- include vehicle safety in employee induction training
- driver training and refresher training given by competent persons
- the provision of driver protection by the use of:
  - seat belts
  - FOPS (fall over protection system)
  - ROPS or TOPS (roll or tip over protection systems)
  - banksmen during reversing operations
- effective vehicular maintenance procedures
- use of daily vehicular inspection checklists
- environmental considerations (road surfaces, gradients and changes in road level, lighting, visibility)

The management of vehicle movements involves:

- a designated management system and a code of practice for all drivers who must be competent
- an effective line management organization to ensure supervision of all vehicular activities
- a documented preventative maintenance programme with all work recorded
- regular inspections and, in some cases, thorough examinations by competent persons
- ensuring that all vehicles fitted with suitable driver protection – seat belts, ROPS
- ensuring that all vehicles fitted with reversing warning systems
- suitable fire precautions, particularly in battery charging areas

Element 10
Manual and Mechanical Handling – Hazards and Control

Learning outcomes

- Describe the hazards and the risk factors which should be considered when assessing risks from manual handling activities
- Suggest ways of minimizing manual handling risk
- Identify the hazards and explain the precautions and procedures to ensure safety in the use of lifting and moving equipment with specific reference to fork-lift trucks, manually operated load moving equipment (sack trucks, pallet trucks), lifts, hoists, conveyors and cranes

Key revision points

- The types of injuries resulting from manual handling operations
- The development of manual handling risk assessments and effective control measures
- Safety in the use of lifting equipment and an understanding of the different types of mechanical handling and lifting equipment
- The legal requirements for the inspection and examination of lifting equipment

Manual handling

Manual handling can involve any load movement by human effort only (lifting, pushing, pulling, carrying or supporting)

Requirements of the manual handling operations regulations are:

- to avoid manual handling, if possible
- to mechanize or automate the lifting process, if possible
- if unavoidable, undertake a risk assessment
- instigate risk controls and review
- give employees information on loads

Manual handling hazards are:	Injuries caused by manual handling include:
➤ lifting a load which is too heavy and/or cumbersome ➤ poor posture and/or poor lifting technique ➤ dropping the load on the foot ➤ lifting sharp-edged or hot loads **Manual handling risk assessment requires:** **The use of mechanical aids; if this is not possible then the assessment of:** ➤ the task ➤ the individual ➤ the load ➤ the working environment	➤ muscular sprains or strains ➤ back injuries ➤ trapped nerve ➤ hernias ➤ cuts, bruises and abrasions ➤ fractures ➤ work related upper limb disorders (WRULDs) ➤ rheumatism

Manual handling training includes:	➤ a good lifting technique, such as:
➤ the types of injury ➤ the manual handling assessment findings ➤ potentially hazardous manual handling operations ➤ the correct use of manually operated load moving equipment, such as sack and pallet trucks ➤ the correct use of mechanical aids ➤ the correct use of personal protective equipment ➤ good housekeeping issues ➤ the factors that can affect an individual	1. Check that suitable clothing (including gloves, if required) and footwear are being worn. Make an approximate assessment of the weight of the load and decide whether to lift it alone. Check to see whether one side of the load is heavier than the other and ensure that the heaviest side of the load is closest to the body 2. Place the feet apart and adopt a good posture by bending the knees and keeping the back straight 3. Get a firm grip and hold the load as close as possible to the body 4. Ensuring that the back remains straight, lift the load to knee level and then to waist level without jerking 5. After ensuring that full visibility is available, move forward without twisting the trunk 6. Set the load down either at waist level or by lowering it first to knee level and then to the floor 7. Manoeuvre the load to its final position after it has been set down

Safety in the use of lifting and moving equipment

Positioning and installation of lifting equipment:	The organization of lifting operations:
Risks during lifting operations should be reduced by avoiding ➤ the equipment or its load from striking a person ➤ a load drifting, falling freely or being released unintentionally ➤ the need to lift loads over people ➤ stopping safely in the event of a power failure • Where possible, enclose the path of the load with a suitable and substantial interlocked gates	Every lifting operation, which is lifting or lowering a load, shall be: ➤ properly planned by a competent person ➤ appropriately supervised ➤ performed in a safe manner **Summary of the requirements for lifting operations** There are four general requirements for all lifting operations: ➤ Use strong, stable and suitable lifting equipment ➤ The equipment should be positioned and installed correctly ➤ The equipment should be visibly marked with the safe working load ➤ Lifting operations must be planned, supervised and performed in a safe manner by competent people

Types of mechanical handling and lifting equipment

<table>
<tr><th>Conveyor</th><th>Elevator</th></tr>
<tr><td>Three types of conveyor – belt, roller and screw

Typical hazards include:
➤ in-running nips
➤ entanglement
➤ loads falling from conveyor
➤ impact against overhead systems
➤ contact hazards – sharp edges
➤ manual handling hazards
➤ noise and vibration hazards</td><td>An elevator transports goods between floors. The most common hazard is caused by loads falling from the elevator. Also there are manual handling hazards at either end of the elevator</td></tr>
</table>

Fork-lift trucks

Hazards include:

- overturning
- overloading
- collisions with pedestrians or other vehicles or structures
- the silent operation of electrically powered fork-lift trucks makes pedestrians unaware of their presence
- uneven road surface
- overhead obstructions
- loss of load
- failure of hydraulic system
- inadequate maintenance
- lack of driver training
- use as a work platform and/or carrying passengers
- speeding
- poor vision around load
- dangerous stacking technique particularly on warehouse racking
- fire – either when battery charging or refuelling

There are also the following physical hazards:

- noise
- exhaust fumes
- vibrations
- manual handling
- ergonomic – musculoskeletal injuries due to soft tyres, expansion joints

Driver daily checks:

- tyres and tyre pressures
- brakes
- reversing horn and light
- mirrors
- secure and properly adjusted seat
- correct fluid levels
- correct working of all lifting and tilting systems

Cranes

Principles of safe operation

- Carry out brief inspection prior to each use (including lifting tackle)
- Do not leave loads suspended when crane not in use
- Prior to lift ensure that nobody can be struck by load or crane
- Never carry loads over people
- Ensure good visibility and communications
- Only lift loads vertically – do not drag load
- Travel with load as close to the ground as possible
- Switch off power to crane when unattended

Mobile jib cranes

- Plan each lift (maximum load and radius of operation)
- Identify overhead obstructions and hazards
- Assess load bearing capacity of ground
- If fitted, outriggers should be used

Reasons for crane failure

- overloading
- poor slinging of load
- overturning
- collision with another structure or overhead power lines
- foundation failure
- structural failure of the crane
- operator error
- lack of maintenance and/or regular inspections

During lifting operations, ensure that:

- driver has good visibility
- there are no pedestrians below load and barriers are in place
- audible warning given prior to lifting operation

Lifts, hoists and items of lifting tackle

A **lift or hoist** incorporates a platform or cage. Its movement is restricted by guides and may carry passengers and/or goods alone.

It is required to:

- be of sound mechanical construction
- have interlocking doors or gates that must be completely closed before the lift or hoist moves
- be fitted, if carrying passengers, with an automatic braking system to prevent overrunning and a safety device to support the lift in the event of suspension rope failure
- be rigorously maintained by competent persons
- be protective of others during maintenance operations from falling down the lift shaft and other hazards

Other **items of lifting tackle** include chain slings and hooks, wire and fibre rope slings, eyebolts and shackles. Important points include:

- Loads must be properly secured and balanced in slings
- Lifting hooks should be checked for wear and hook distortion
- Shackles and eyebolts must be correctly tightened
- Slings must be checked for any damage before use and only used by competent people
- Training and instruction should be given in the use of lifting tackle
- Regular inspections of tackle should be made in addition to the mandatory thorough examinations
- Lifting items should be carefully stored between use

The equipment should be inspected at suitable intervals between thorough examinations. The frequency and the extent of the inspection is determined by the level of risk presented by the lifting equipment. A report or record should be made of the inspection, which should be kept until the next inspection. Unless stated otherwise, lifts and hoists should be inspected every week.

Statutory examination of lifting equipment

The Lifting Operations and Lifting Equipment Regulations (LOLER) require inspections and thorough examinations.

An **inspection** is performed by a competent person appointed by the employer and is used to identify whether the equipment can be operated, adjusted and maintained safely.

The following points apply to inspections:

- They should take place between thorough examinations
- The frequency depends on risk assessment
- The report or record should be kept until the next inspection
- Lifts and hoists should be inspected every week

A **thorough examination** is a detailed examination of lifting equipment by a competent person who is independent of the employer. The examination is usually made in accordance with a written scheme and a report (within 28 days) is submitted to the employer. The report must be kept for 2 years or until the next thorough examination (whichever is longest).

A thorough examination should be undertaken at the following times:

- before first use
- after it has been assembled at a new location
- at least every 6 months for lifting equipment or lifting accessory used for lifting persons
- at least every 12 months for all other lifting equipment
- in accordance with a particular examination scheme drawn up by an independent competent person
- each time that exceptional circumstances, which are likely to jeopardize the safety of the lifting equipment, have occurred (such as severe weather)

Element 11
Work Equipment Hazards and Control

Learning outcomes

- Outline general requirements for work equipment
- Outline the hazards and controls for hand tools
- Describe the main mechanical and non-mechanical hazards of machinery
- Describe the main methods of protection from machinery hazards

Key revision points

- The legal requirements for the supply and use of work equipment
- The safe use and maintenance of work equipment
- The hazards and controls related to hand-held tools
- Examples of mechanical and non-mechanical machinery hazards
- Practical safeguards and their application to a range of machines
- The construction requirements of guards

There are TWO types of law affecting work equipment:

- equipment supply law – designed to prevent barriers to trade – The Supply of Machinery (Safety) Regulations, and
- equipment user law – designed to protect people at work – The Provision and Use of Work Equipment Regulations

Supply of work equipment

Most new work equipment should have 'CE' marking when purchased. The 'CE' mark requires manufacturers to use the following hierarchy:

- Identify the health and safety hazards likely to be present when the machine is used
- Assess the associated risks
- Remove the significant hazards at the design stage or, if that is not possible
- Provide safeguards (such as guarding or noise enclosures) or, if that is not possible
- Use warning signs on the machine to warn of hazards

Additional requirements of equipment manufacturers:

- Retain related information in a technical file
- Fix a CE mark to the equipment – indicates compliance with all relevant supply laws
- Issue a 'Declaration of Conformity' for the equipment

The Declaration of Conformity states that the equipment complies with all relevant health and safety requirements. The Declaration must:

- state the name and address of the manufacturer or importer into the EU
- contain a description of the equipment, and its make, type and serial number
- indicate all relevant European Directives with which the machinery complies
- state details of any notified body that has been involved
- specify which standards have been used in the manufacture (if any)
- be signed by a person with authority to do so
- provide the purchaser with instructions on the safe installation, use and maintenance of the equipment

Use of work equipment

Definition of work equipment is: any equipment used by an employee at work, such as:	Uses of work equipment include:
➤ hand tools – hammers, knives, spanners ➤ machines – photocopiers, drilling machines, vehicles ➤ apparatus – laboratory equipment ➤ lifting equipment – fork-lift trucks, hoists ➤ other equipment – ladders, vacuum cleaners	➤ starting and stopping ➤ repair and maintenance ➤ modification and servicing ➤ cleaning ➤ transporting

General requirements of work equipment

Management duties

The equipment must be:

- suitable for its purpose of use and only used for specified operations
- suitable in terms of initial integrity and place of use
- selected with the health and safety of the user in mind
- maintained in an efficient working order and in good repair
- inspected at regular intervals
- restricted in use to designated persons when it has specific hazards
- used together with appropriate information, instruction and training
- conformed to EC requirements

Other duties

This includes:

- the guarding of dangerous parts of work equipment
- the provision of stop and emergency stop controls
- ensuring that the equipment is stable
- ensuring the lighting around the equipment is suitable and sufficient
- ensuring suitable warning markings or devices, such as flashing lights, are fitted

Hierarchy of risk control for work equipment

- Eliminate the risks
- Control risks by physical methods (such as guarding)
- Implement software measures (such as safe systems of work)

Information, instruction and training

This should include:

- the significant health and safety issues with the equipment
- any limitations or problems with the use of the equipment
- safe methods to deal with these problems
- details of any residual risks associated with the equipment
- details of safe working procedures
- the location of the manufacturer's manual or guidance

Induction training will be required on recruitment but refresher training will also be needed when:

- jobs are changed, particularly if the level of risk increases
- new technology or equipment is introduced
- the system of work is modified
- there are legislative changes
- a periodic refresher course is due

Types of maintenance

- preventative planned maintenance
- condition based maintenance
- breakdown based maintenance

Hazards associated with maintenance

- lack of competence and/or training
- equipment not made safe before maintenance
- lack of permit to work
- use of incorrect tools or unsafe equipment

Inspection of work equipment

- after installation or put into service for the first time
- after assembly at a new site or in a new location
- at suitable intervals
- each time exceptional circumstances occur that could affect safety

Statutory requirements for examination of boilers and air receivers

- supplied with correct written information and markings
- properly installed
- used within their operating limits
- written scheme for periodic examination certified by a competent person
- examined by a competent person in accordance with the written scheme
- a report of the periodic examination held on file and any required actions undertaken

Equipment controls

The equipment should have:

- operating controls – easily reached, operated and with adequate markings and warning signs
- no obstructions or debris around it
- emergency controls (red emergency stop buttons)
- stability during use

Other issues

- the level and quality of general and local lighting
- clear and durable markings on equipment

Hand-held tools

Typical hazards include:	Hand tool controls
➤ broken handles ➤ incorrect use of knives, saws and chisels ➤ tools that slip ➤ splayed spanners ➤ flying particles ➤ electrocution and/or burns	➤ .suitable for purpose – (e.g. insulated tools for electricians) ➤ .inspection on a regular basis ➤ .training for all hand tool users

Hand-held power tools

Hazards include:

- entanglement
- flying particles – (eye injury)
- contact with cutting edges
- striking hidden/buried services (electrical cables, gas mains)
- manual handling injuries
- hand–arm vibration syndrome
- noise levels
- dust levels
- trailing leads
- fire and/or explosion

Typical controls:

- Read operating instructions before initial use
- Maintain a clean, tidy and well lit working area
- Never use power tools near water, combustible fluids or gases
- Store tools in a dry, safe and secure place
- Never overload tools
- Always use the correct tool and/or attachment for heavy job
- Always wear suitable work clothes and appropriate personal protective equipment
- Ensure that the work piece is securely held in clamps or a vice
- Maintain tools with care, keep them clean and sharp and inspect cables regularly
- Always check that the power switch is turned off before connecting the power cable
- Use only tool accessories and attachments recommended by the tool manufacturer

Mechanical machinery hazards are:	**Non-mechanical machinery hazards include:**
➤ crushing ➤ shearing ➤ cutting or severing ➤ entanglement ➤ drawing-in or trapping – (in-running nips) ➤ impact ➤ stabbing or puncture ➤ friction or abrasion ➤ high pressure fluid ejection	➤ access to the machinery ➤ slips, trips and falls ➤ objects falling and moving from the machinery ➤ obstructions and projections ➤ manual handling and lifting ➤ electricity (shock, burns or fire) ➤ fire and explosion ➤ noise and vibration ➤ high pressure – ejection from hydraulic hose leaks ➤ high/low temperature ➤ dust, fume or mist ➤ radiation – ionising and non-ionising

Examples of equipment with machinery hazards

- photocopier
- document shredder
- bench top grinding machine
- pedestal drill
- cylinder mower
- strimmer/Brush-cutter
- chainsaw
- compactor
- checkout conveyor system
- cement mixer
- bench-mounted circular saw

Practical safeguarding

Defined by the legal definition of 'practical' – if it is technically possible to guard an item of equipment then it should be guarded irrespective of cost and convenience.

Levels of protection or hierarchy of measures

Regulation 11 of PUWER

1. fixed enclosing guarding
2. other guards or protection devices (interlocked, trip, etc.)
3. protection appliances (jigs, push-sticks, etc.)
4. provision of information, instruction, training and supervision

Guard	Example
Fixed	Around a fan
Adjustable	Around a drill bit on a pedestal drill
Interlocking	On the gates of a passenger lift, microwave oven
Trip device	Photoelectric cell in front of a guillotine
Two-handed control	Crimping machine
Hold-to-run	Tube train driver

Types of guard

Apply the above to the 11 pieces of equipment listed earlier – photocopier, document shredder, bench top grinding machine, pedestal drill, cylinder mower, strimmer/brush-cutter, chain saw, compactor, checkout conveyor system, cement mixer and a bench-mounted circular saw.

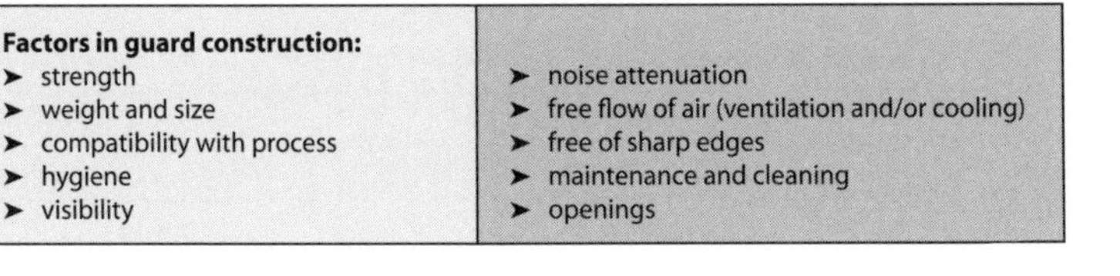

Factors in guard construction:

- strength
- weight and size
- compatibility with process
- hygiene
- visibility
- noise attenuation
- free flow of air (ventilation and/or cooling)
- free of sharp edges
- maintenance and cleaning
- openings

Element 12
Electrical Hazards and Control

Learning outcomes

- Identify the hazards and evaluate the consequential risks from the use of electricity in the workplace
- Describe the control measures that should be taken when working with electrical systems or using electrical equipment

Key revision points

- The basic electrical principles and definitions
- The causes and treatment of electric shock
- Other electrical hazards and injuries and their associated controls
- The selection, suitability, maintenance and inspection of electrical equipment
- The application, advantages and limitations of portable electrical appliance testing

Electrical principles and definitions:	**Electrical hazards and injuries:**
➤ voltage (volts V) ➤ current (amps A) ➤ resistance (ohms Ω) ➤ Ohm's law – V = I x R ➤ Power P = V x I watts ➤ conductors – electricity will flow easily ➤ insulators – very poor conductors ➤ earthing – connection to earth ➤ short circuit – direct flow to earth ➤ low voltage – not exceeding 600 V ➤ high voltage – exceeding 600 V ➤ mains voltage – 220/240 V	➤ electric shock ➤ electric burns ➤ electric fires and explosions ➤ arcing ➤ portable electrical equipment ➤ secondary hazards (falls, trips, noise and vibration)

Electric shock

- The effect of an electric shock can vary from a slight tingling sensation to death
- The severity of the shock depends on the size of the current and voltage
- Low voltage shocks, below 110V, are seldom fatal
- Electric shock can also produce burns

Treatment of electric shock

- Raise the alarm
- Switch off the power – if not possible, use insulating material to move victim from contact with the power supply
- Call for an ambulance
- If breathing, place victim in recovery position
- If not breathing, apply mouth-to-mouth resuscitation
- Treat any burns
- If person regains consciousness, treat for normal shock
- Remain with person until help arrives

If high voltages are involved, inform police and electrical supply company but do not approach the victim

Electrical fires and explosions

25% of all fires have an electrical origin

Causes

- short circuits
- overheating of cables and equipment
- ignition of flammable gases and vapours
- ignition of combustible substances by static electrical discharges
- electric arcing
- static electricity – lightning strikes

Portable electrical equipment

25% of all electrical accidents are caused by portable electrical equipment

Hazards

- faulty cables, extension leads, plugs and sockets
- inadequate maintenance
- don't use in flammable or damp atmospheres
- misuse of equipment

General control measures for electrical hazards

The main control measures are given in the Electricity at Work Regulations which cover the following topics:

- the design, construction and maintenance of electrical systems, work activities and protective equipment
- the strength and capability of electrical equipment
- the protection of equipment against adverse and hazardous environments
- the insulation, protection and placing of electrical conductors
- the earthing of conductors and other suitable precautions
- the integrity of referenced conductors
- the suitability of joints and connections used in electrical systems
- means for protection from excess current
- means for cutting off the supply and for isolation
- the precautions to be taken for work on equipment made dead
- working on or near live conductors
- adequate working space, access and lighting
- the competence requirements for persons working on electrical equipment to prevent danger and injury

Specific controls for electrical hazards

- a management system that ensures safe installation, operation and maintenance
- training at all levels – induction, supervisory and technical
- safe systems of work, including permits to work on live electricity
- selection of suitable equipment for the work and environment by considering the following:
  - atmosphere (flammable, damp)
  - weather conditions
  - high or low temperatures
  - dirty or corrosive conditions
  - standard of installation
  - rated operating conditions
- the use of protective systems
- inspection and maintenance strategies

Protective systems

- fuse (and circuit breaker)
- insulation
- isolation
- reduced low voltage systems
- residual current devices
- double insulation

Inspection and maintenance strategies

The particular areas of interest for inspection and maintenance are:

- the competence of the inspection and maintenance team
- the cleanliness of insulator and conductor surfaces
- the mechanical and electrical integrity of all joints and connections
- the integrity of mechanical mechanisms, such as switches and relays
- the calibration, condition and operation of all protection equipment, such as circuit breakers, RCDs and switches
- isolation procedures
- inspection and maintenance records
- permits-to-work for work on live electricity

Inspection strategies

Any strategy for the inspection of electrical equipment, particularly portable appliances, should involve the following considerations:

- a means of identifying the equipment to be tested
- the number and type of appliances to be tested
- the competence of those who will undertake the testing (whether in-house or bought-in)
- the legal requirements for portable appliance testing (PAT) and other electrical equipment testing and the guidance available
- organizational duties of those with responsibilities for PAT and other electrical equipment testing
- test equipment selection and re-calibration
- the development of a recording, monitoring and review system
- the development of any training requirements resulting from the test programme

Portable electrical appliance testing

- user checks – at least once a week
- formal visual inspection – by a trained person
- combined testing and inspection – by a competent person using testing equipment
- the frequency of inspection and testing depends on usage rate and conditions – normally determined by a risk assessment
- records of inspection and testing required

Advantages of portable appliance testing

- early indication of serious faults
- discovery of incorrect equipment or electrical supply
- incorrect fuses found
- reduction in the number of electrical accidents
- monitoring misuse of portable appliances
- check on selection procedures
- increase awareness of hazards
- introduction of a more regular maintenance regime

Limitations of portable appliance testing

- too frequent testing of fixed equipment – can lead to excessive costs
- some unauthorized equipment never tested because of lack of records
- equipment misused or overused between tests
- the inclusion of trivial faults – leads to long lists and the overlooking of significant faults
- the level of competence of tester
- the use of incorrectly calibrated test equipment

Element 13
Fire Hazards and Control

Learning outcomes

- Identify fire hazards and evaluate main fire risks in a workplace
- Explain the basic principles of fire prevention and the prevention of fire spread in buildings
- Identify the appropriate fire alarm system and fire-fighting equipment for a simple workplace
- Outline the requirements for an adequate and properly maintained means of escape for a simple workplace
- Outline the factors which should be considered when implementing a successful evacuation of a workplace in the event of a fire

Key revision points

- The legal framework for the control of fire hazards
- The basic principles of fire prevention – classification, methods of extinction and principles of heat transmission
- The common causes of fire, its spread and consequences
- Key elements of a fire risk assessment
- The principles of fire detection, protection, warning and means of escape
- The provision, maintenance and testing of fire fighting equipment
- Emergency planning and training

Methods of extinction

- cooling – reducing ignition temperature
- smothering – limiting oxygen available
- starving – limiting fuel supply
- chemical reaction – interrupting the chain of combustion

Basic principles

The fire triangle consists of:

- fuel – solids, liquids and gases
- ignition source
- oxygen – from air, oxidizing materials

Sources of ignition

- naked flames
- external sparking
- internal sparking
- hot surfaces
- static electricity

Classes of fire

- A – carboniferous solids
- B1 – liquids soluble in water
- B2 – liquids insoluble in water
- C – gases
- D – metals
- F – high temperature cooking oils or fats

Electrical fires do not constitute a separate fire class.

Principles of heat transmission and fire spread

- convection – through air movements (floor to floor)
- conduction – through conducting materials (girders)
- radiation – emission of heat waves (bar heaters)
- direct burning – contact with flames (rubbish catching fire)

Causes of fire:

- rubbish (particularly if burnt)
- electrical short circuits
- portable heaters
- friction causing sparks (hand grinders)
- oil heating plant
- cooking appliances
- arson (often by children)
- hot surfaces (welding)
- cigarette smoking

Fire spreads via:

- gases/sprays
- foam plastics
- packaging
- previous spillage
- flammable dusts

Fire spreads due to:

- delayed discovery
- alarm not raised
- fire service not informed
- lack of barriers
- fire procedures not known
- fire extinguishers unavailable, not serviced or not used
- air flow – particularly if fire doors not closed
- lack of or poor emergency lighting
- flammable materials
- flammable dusts

Fire extinguishers

Fire extinguishers are coloured red over 95% of surface

Colour (5%)	Contents	Class/type of fire
Red	Water	A
Cream	AFFF	A (possibly B)
Cream	Foam	B and C
Black	Carbon dioxide	B and electrical
Blue	Dry powder	B, C and electrical

Fire risk assessment

Required by the Regulatory Reform (Fire Safety) Order 2005

Stages of the fire risk assessment

- Identify the fire hazards – identify sources of heat/ignition, unsafe acts (smoking) and unsafe conditions (storage of combustible materials)
- Identify all locations and persons at risk
- Reduce the risks by the introduction of controls
- Record the findings
- Monitor and review

Hazards

- sources of fuel
- oxygen depletion
- flames and heat
- smoke
- gaseous combustion products
- structural failure of the building

Other factors

- time to detect and warn
- safe egress
- adequacy of fire signs
- provision of fire fighting equipment
- fire training – fire drills, appliance use and emergency teams
- maintenance of fire equipment
- consultation with employees

Control measures

- elimination or reduction in the use and storage of flammable materials
- control of ignition sources
- effective systems of work
- good housekeeping
- storage facilities for small quantities (50 litres) of highly flammable or flammable liquids

Maintenance and testing of fire fighting equipment

Fire detection and warning

- alarms – including smoke and heat detectors
- audibility throughout premises
- regularly tested
- records of all tests
- vigilance of workers following training

Means of escape

- doors – clear of obstructions and opening in direction of escape
- escape routes (stairways and passageways) directly to a place of safety
- lifts should not be used
- lighting including emergency lighting
- directional and other signs
- escape times
- assembly points

Structural design measures

- fire loading of the building
- surface spread of fire
- fire resistance of structural elements
- insulating materials
- fire compartmentation

Emergency planning and training

- evacuation procedures
- appointment of fire marshals
- fire drills
- roll call
- provision for the infirm and disabled

Fire procedures for people with a disability

Emergency planning needs to consider the special requirements of people with physical or mental disabilities. Such issues include:

- the identification of those who may need special help to evacuate the building
- the allocation of responsibility to specific staff to identify and assist people with a disability during emergencies
- planning suitable escape routes
- having a system in place that enables a person with a disability to summon help in emergencies
- including the arrangements for disabled persons in all staff fire training programmes

Element 14
Chemical and Biological Hazards and Control

Learning outcomes

- Recognize the forms of, and classification of, substances hazardous to health
- Explain the factors to be considered when undertaking a preliminary assessment of the health risks from substances commonly encountered in the workplace
- Describe the use and limitations of Workplace Exposure Limits including the purpose of long-term and short-term exposure limits
- Distinguish between acute and chronic health effects
- Outline control measures that should be used to reduce the risk of ill-health from exposure to hazardous substances
- Outline the basic requirements related to disposal of waste and effluent (and the control of atmospheric pollution)

Key revision points

- The details and classification of hazardous chemical and biological agents
- The functions and vulnerability of various human body organs to hazardous substances
- The requirements of the COSHH Regulations – assessment and recommended control measures
- Health surveillance and maintenance controls
- Emergency controls and environmental issues

Agents

Chemical agents:	Biological agents:
➤ dusts – including respirable dust (fine dust that remains in the lungs) ➤ gases ➤ vapours – substances very close to their boiling points (e.g. solvents) ➤ liquids ➤ mists – similar to vapours but closer to the liquid phase (e.g. paint sprays) ➤ fume – condensed metallic particles (e.g. welding fume)	➤ fungi ➤ moulds ➤ bacteria ➤ viruses

Classification of hazardous substances

- Irritant – repeated contact with skin can cause a sensitized or allergic inflammation
- Harmful – involve limited health risks
- Corrosive – may destroy living tissue
- Toxic – produce serious health risks
- Infectious – contain micro-organisms known to cause diseases
- Carcinogenic – may induce cancer
- Mutagenic – may induce hereditary genetic defects

Acute effects are of short duration, normally reversible and appear shortly after exposure to a hazardous substance (e.g. asthma attacks).
Chronic effects develop over a period of time after repeated exposure to a hazardous substance and are often irreversible.

Major human body systems

There are 5 major functional systems within the body. These are shown together with typical diseases and related hazardous substances

Routes of entry into the body are:

- ➤ inhalation via the lungs
- ➤ absorption via the skin
- ➤ ingestion via the stomach
- ➤ injection through the skin

Tumours may be benign or malignant.

Body system	Illness or disease	Hazardous substance
Respiratory	Bronchitis, asthma, fibrosis, cancer	Dusts, asbestos
Nervous	Anxiety, epilepsy, loss of consciousness	Solvents, lead
Cardiovascular	Headaches, loss of consciousness	Benzene, carbon monoxide
Urinary	Cirrhosis, cancer	Chlorinated hydrocarbons, mercury
Skin	Irritant contact dermatitis and allergic contact dermatitis	Detergents, turpentine, epoxy resin

Health hazards of specific agents

Each agent is followed by a brief description and two possible illnesses

Ammonia – corrosive gas can cause burns to skin and eyes and severe bronchitis
Chlorine – toxic, irritant gas can cause nausea and severe bronchitis
Organic solvents – sensitizer, irritant can cause dermatitis and liver failure
Carbon dioxide – odourless gas can cause headaches and loss of consciousness
Carbon monoxide – odourless gas can cause headaches and unconsciousness
Isocyanates – sensitizer, irritant can cause bronchitis and extreme 'asthma attack'
Asbestos – fine fibrous dust can cause fibrosis and mesothelioma
Lead – dust can cause headaches, nausea and anaemia
Silica – dust can cause fibrosis and silicosis
Leptospira – bacteria in rat urine can cause anaemia and jaundice
Legionella – bacteria in tepid water can cause pneumonia and death
Hepatitis – from a virus or organic solvents can cause nausea and jaundice

Note: Any potential source of legionella (e.g. cooling towers) must be given a special risk assessment by a competent person and records kept.

Control of Substances Hazardous to Health Regulations

Requirements of COSHH

- Suitable and sufficient health risk assessment
- Adequate control of exposure of employees
- Proper use of control measures provided
- Maintain control measures
- Monitor employees exposed to hazardous substances (Schedule 5)
- Health surveillance (Schedule 6)
- Information, instruction and training

COSHH assessment

- Identify hazardous substances
- Gather information
- Evaluate the risks
- Decide on control measures
- Record the assessment
- Review the assessment

Occupational exposure limits

These are called workplace exposure limits (WELs) and must not be exceeded.

Two categories – (similar to previous MEL and OES):

- Category 1 – carcinogenic and mutagenic substances – exposure must be reduced as low as is reasonably practicable below the WEL
- Category 2 – all other hazardous substances – exposure controlled by principles of good practice

There are two time weighted averages – TWA:

- the long-term exposure limit (LTEL) – over 8 hours
- the short-term exposure limit (STEL) – 15 minute reference period

Sources of information

- product labels
- safety data sheets
- HSE Guidance Note EH 40
- trade association publications
- industrial codes of practice
- specialist reference manuals

Survey techniques for health risks

Survey techniques for health risks include:

- Stain tube detector – The advantages of this technique are that it is quick, relatively simple to use and inexpensive. The disadvantages are:
  - it cannot be used to measure concentrations of dust or fume
  - the accuracy of the reading is approximately $\pm 25\%$
  - it will yield false readings if other contaminants present react with the crystals
  - it can only give an instantaneous reading not an average reading over the working period (TWA)
  - the tubes are very fragile with a limited shelf life
- Passive sampling including the static dust sampler
- Sampling pumps and heads
- Direct reading instruments (MIRA)
- Vane anemometers and hygrometers
- Smoke tube and dust observation lamp

Control measures under COSHH

Principles of good practice for the control of exposure to substances hazardous to health

1. Design and operate processes and activities to minimize emission, release and spread of substances hazardous to health
2. Take into account all relevant routes of exposure – inhalation, skin absorption and ingestion – when developing control measures
3. Control exposure by measures that are proportionate to the health risk
4. Choose the most effective and reliable control options which minimize the escape and spread of substances hazardous to health
5. Where adequate control of exposure cannot be achieved by other means, provide, in combination with other control measures, suitable personal protective equipment
6. Check and review regularly all elements of control measures for their continuing effectiveness
7. Inform and train all employees on the hazards and risks from the substances with which they work and the use of control measures developed to minimize the risks
8. Ensure that the introduction of control measures does not increase the overall risk to health and safety

Hierarchy of control measures

Measures for preventing or controlling exposure to hazardous substances include one or a combination of the following:

- elimination of the substance
- substitution of the substance (or reduction in the quantity used)
- total or partial enclosure of the process
- local exhaust ventilation
- dilute or general ventilation
- reduction of the number of employees exposed to a strict minimum
- reduced time exposure by task rotation and the provision of adequate breaks
- good housekeeping
- training and information on the risks involved
- effective supervision to ensure that the control measures are being followed
- personal protective equipment (such as clothing, gloves and masks)
- welfare (including first aid)
- medical records
- health surveillance

Preventative control measures

Engineering controls

Engineering controls include:

- segregation of people from the process (fume cupboard)
- modification of the process to reduce human contact with the hazardous substances
- local exhaust ventilation comprising:
  - a collection hood and intake
  - a ventilation duct
  - a filter or other air cleaning device
  - a fan
  - an exhaust duct
- dilution or general ventilation

Supervisory or people controls

Additional supervisory controls when hazardous substances are involved are:

- reduced time exposure – thus ensuring that workers have breaks in their exposure periods
- reduced number of workers exposed – only persons essential to the process should be allowed in the vicinity of the hazardous substance. Walkways and other traffic routes should avoid any area where hazardous substances are in use
- eating, drinking and smoking – these must be prohibited in areas where hazardous substances are in use
- any special rules, such as the use of personal protective equipment, strictly enforced

Personal protective equipment

Personal protective equipment (PPE) is the control measure of *last resort*. It is covered by the Personal Protective Equipment at Work Regulations which require:

- PPE to be suitable for wearer and task
- PPE to be compatible with any other PPE provided
- a risk assessment to be undertaken to determine best PPE
- a suitable PPE maintenance programme
- suitable storage arrangements for the PPE
- information, instruction and training for the user of the PPE
- supervision of the use of PPE
- a system for reporting defects

Note: the Regulations do not cover Respiratory protective equipment (RPE)

Types of personal protective equipment	**Respiratory protective equipment**
➤ hand and skin protection ➤ eye protection ➤ protective clothing – boots, hard hats, aprons	➤ filtering half mask ➤ half mask respirator ➤ full face mask respirator ➤ powered respirator

Health surveillance

Health surveillance is not same as health monitoring

Types of health surveillance include:

- checking individuals on a regular basis
- checking individuals as a result of statistical evidence of a potential problem
- when a substance listed in Schedule 6 of COSHH is being used. In these cases, health surveillance should take place at intervals not exceeding 12 months and records kept for 40 years

Maintenance and emergency controls

Maintenance involves:

- the cleaning, testing and, possibly, the dismantling of equipment
- the changing of filters in extraction plant or entering confined spaces
- the handling of hazardous substances and the safe disposal of waste material
- a permit-to-work procedure to be in place (this may be required) since the control equipment will be inoperative during the maintenance operations
- the keeping of records of maintenance for at least 5 years

The following points should be considered when emergency procedures are being developed:

- the possible results of a loss of control (e.g. lack of ventilation)
- dealing with spillages and leakages (availability of effective absorbent materials)
- raising the alarm for more serious emergencies
- evacuation procedures including the alerting of neighbours
- fire fighting procedures and organization
- availability of respiratory protection equipment
- information and training

Environment considerations

The legal framework is defined by the Environmental Protection Act and the Pollution, Prevention and Control Act

Air pollution	Water pollution
➤ Part A processes – air, water and land ➤ Part B processes – air only ➤ Prescribed processes and substances	➤ Discharges to sewers ➤ Bund wall around oil store

Waste disposal

- Environmental permits
- Authorized persons for waste disposal
- Packaging Regulations
- Hazardous Waste Regulations – (Special Waste Regulations still in Scotland)

Element 15
Physical and Psychological Health Hazards and Control

Learning outcomes

- Identify work processes and practices that may give rise to musculoskeletal health problems (in particular, work-related upper limb disorders – WRULDs) and suggest practical control measures
- Identify common welfare and work environment requirements in the workplace
- Describe the health effects associated with exposure to noise and suggest appropriate control measures
- Describe the health effects associated with exposure to vibration and suggest appropriate control measures
- Describe the principal health effects associated with ionising and non-ionising radiation and outline basic protection techniques
- Explain the causes and effects of stress at work and suggest appropriate control actions
- Describe the situations that present a risk of violence towards employees and suggest ways of minimizing such risk

Key revision points

- Musculoskeletal health issues and ill-health prevention strategies
- The legal requirements for vibration and noise control, display screen use, workplace safety and ionising radiations
- Welfare and work environment requirements
- The hazards and controls associated with ionising and non-ionising radiations
- The causes and prevention of workplace stress and violence

Task and workstation design

Ergonomics – interaction between the worker, their work and their environment. It involves knowledge of their physical and mental capabilities in addition to their understanding of the job

Ill-health effects due to poor ergonomics

Work-related upper limb disorders (WRULDs):

- tenosynovitis
- repetitive strain injury (RSI)
- carpal tunnel syndrome
- frozen shoulder

Control of vibration at work regulations

Preventative and control measures for:

- hand–arm vibration (HAVS)
- whole body vibration (WBV)

Ill health due to hand–arm vibrations:	Ill health due to whole body vibration:
➤ Hand–arm vibration syndrome (HAVS) ➤ Vibration white finger (VWF) ➤ Carpal tunnel syndrome	➤ Severe back pain ➤ Reduced visual and manual control ➤ Increased heart rate and blood pressure ➤ Spinal damage

Preventative and precautionary measures

The common measures used to control ergonomic ill-health effects are:

- Identify repetitive actions
- Eliminate vibration by performing the job in a different way
- Ensure that the correct equipment (properly adjusted) is always used
- Introduce job rotation so that workers have a reduced time exposure to the hazard
- During the design of the job ensure that poor posture is avoided
- Undertake a risk assessment
- Examine ill-health reports and absence records
- Introduce a programme of health surveillance
- Ensure that workers are given adequate information on the hazards and develop a suitable training programme
- Ensure that a programme of preventative maintenance is introduced and include the regular inspection of items such as vibration isolation mountings
- Keep up to date with advice from equipment manufacturers, trade associations and health and safety sources

Display screen equipment (DSE)

Requirements of the Health and Safety (Display Screen Equipment) Regulations:

- suitable and sufficient risk assessment
- workstation compliance with minimum specifications given in the Regulations
- adequate breaks in work programme
- free eye sight test if required
- information and training

Ill-health hazards

- musculoskeletal problems
- visual problems
- psychological problems

DSE risk assessment

The DSE risk assessment should consider the following factors:

- the height and adjustability of the monitor
- the adjustability of the keyboard, the suitability of the mouse and the provision of wrist support
- the stability and adjustability of the DSE user's chair
- the provision of ample foot room and suitable foot support
- the effect of any lighting and window glare at the work station
- the storage of materials around the work station
- the safety of trailing cables, plugs and sockets
- environmental issues – noise, heating, humidity and draughts

Welfare and work environment issues

Welfare and work environment issues are covered by the Workplace (Health, Safety and Welfare) Regulations.

Welfare

- sanitary conveniences and washing facilities
- drinking water
- clothing and changing facilities
- rest and eating facilities

Workplace environment

- ventilation
- heating and temperature
- lighting
  - natural light
  - stroboscopic effects
  - special local lighting
  - structural aspects (shadows)
  - atmospheric dust
  - heating effect of lighting
  - lamp and window cleaning
  - emergency lighting
- workstations and seating
- floors, stairways and traffic routes
- translucent or transparent doors constructed with safety glass and properly marked to warn pedestrians of their presence
- adequate arrangements in place to ensure the safe cleaning of windows and skylights
- adequate provisions for the needs of disabled workers

Noise

Requirements of Noise at Work Regulations

- Assess noise levels and keep records
- Reduce the risks from noise exposure by engineering controls – only use hearing protection as a last resort
- Provide employees with information and training
- If a manufacturer of equipment, to provide information on noise levels of the equipment

Health effects of noise

- Human ear
  - ear drum
  - cochlea
  - hair cells – damage irreversible
  - auditory nerve
- Acute effects
  - temporary threshold shift
  - tinnitus
  - acute acoustic trauma
- Chronic
  - noise-induced hearing loss
  - permanent threshold shift
  - tinnitus

Noise assessments

Objective is to identify whether action levels have been reached

Noise measurement

- sound pressure level (SPL) – dB(A)
- peak sound pressure
- continuous equivalent noise level (L_{Eq}) – normally measured over an 8-hour period
- daily personal exposure level ($L_{EP,d}$)

Noise action levels

- first action level – daily exposure level of 80 dB(A) – action advised
- second action level – daily exposure level of 85 dB(A) – action obligatory
- peak action level – 135 dB(C) and 137 dB(C)
- exposure limit values 87 dB(A) and 140 dB(C)

Noise control techniques include:

- Reduction of noise at source
  - change equipment or process
  - change speed
  - improve maintenance
- Re-location of equipment
- Enclosing equipment
- Screens or absorption walls
- Damping
- Lagging
- Silencers
- Isolation of workers
- Suitable warning signs

Personal ear protection

- ear plugs
- ear defenders (earmuffs)

Factors to be considered for personal ear protection:

- suitability (frequencies)
- acceptability and comfort
- durability
- instruction in use
- hygiene considerations
- beards, hair and spectacles may reduce effectiveness of protection
- maintenance and storage
- cost

Heat and radiation hazards

Extremes of temperature

There are health problems at both high and low temperatures for the body.

Ionising radiation

Ionising radiation is produced by alpha, beta and gamma particles.

Harmful effects

- Somatic – cell damage to the individual
- Genetic – cell damage to the children of the individual
- Acute – nausea, vomiting, skin burns and blistering, collapse and death
- Chronic – anaemia. leukaemia, other types of cancer

Sources

- radon gas
- X-ray equipment
- smoke detectors

Non-ionising radiation

- Ultraviolet radiation – sun, arc welding
- Lasers – eye and skin burns, electricity, etc.
- Infra-red radiation – fires, furnaces, etc.
- Microwaves – cookers/ovens, mobile telephones

Radiation protection strategies

Ionising radiation

Covered by the Ionising Radiations Regulations which specify:

- risk assessment
- shielding
- time (reduced time exposure)
- distance
- training
- personal protective equipment
- no food or drink consumption near exposed areas
- signs and information
- medical surveillance
- maintenance and inspection controls
- emergency procedures
- radiation protection supervisor – appointed to advise on compliance with Regulations
- radiation protection adviser – appointed to advise the radiation protection supervisor

Non-ionising radiation

- eye protection
- skin protection (gloves and/or creams)
- fixed shields and non-reflective surfaces
- interlocking guards on microwaves

Workplace stress

The reaction of the body to excessive mental pressure that may lead to ill health.

Causes

- The job itself – unrealistic targets, boring, repetitive, insufficient training
- Individual responsibility – ill-defined roles, too much responsibility, too little control to influence outcome
- Working conditions – lack of privacy or security, unsafe practices, threat of violence, excessive noise
- Management attitudes – negative health and safety culture, poor communication, lack of support in a crisis
- Relationships with colleagues – bullying, harassment

Prevention

- Take a positive attitude to stress
- Take employee concerns seriously
- Use effective communication and consultation
- Develop a policy on stress
- Provide relevant training
- Employee appraisal system
- Discourage excessive hours at work
- Encourage life style changes
- Monitor incidents of bullying, etc.
- Avoid blame culture
- Set up a confidential counselling advice service

Workplace violence

Any incident in which a person is abused, threatened or assaulted in circumstances related to their work. High risk occupations include health and social services, police and fire fighters, lone and night workers and benefit services.

Action plan

1. Find out if there is a problem
2. Decide on what action to take:
   i. quality of service provided
   ii. design of the operating environment
   iii. type of equipment used
   iv. designing the job
3. Take the appropriate action
4. Check that the action is effective

Element 16
Construction Activities – Hazards and Control

Learning outcomes

- Identify the main hazards of construction and demolition work and outline the general requirements necessary to control them
- Identify the hazards of work above ground level, outline the general requirements necessary to control them and describe the safe working practices for common forms of access equipment
- Identify the hazards of excavations and outline the general requirements necessary to control them
- Identify the hazards to health commonly encountered in small construction activities and explain how risks might be reduced

Key revision points

- The scope of construction and the associated hazards and controls
- The health and safety management of construction activities and the legal duties under the Construction (Design and Management) Regulations
- The issues involved with working at height
- The hazards and controls required for excavation work

The scope of construction

The scope of the construction industry includes:

- general building work – domestic, commercial or industrial may be:
  - new building work, such as a building extension, or
  - the refurbishment, maintenance or repair of existing buildings
- civil engineering projects involving road and bridge building, water supply and sewage schemes and river and canal work
- the use of woodworking workshops together with woodworking machines and their associated hazards
- electrical installation and plumbing work
- painting and decorating
- work in confined spaces such as excavations and underground chambers

Construction hazards and controls

1. Safe place of work

- secure and locked gates with appropriate notices posted
- a secure and undamaged perimeter fence with appropriate notices posted
- all ladders either stored securely or boarded across their rungs
- all excavations covered
- all mobile plant immobilized, where practicable, and services isolated
- secure storage of all inflammable and hazardous substances
- visits to local schools to explain the dangers present on a construction site
- if unauthorized entry persists, then security patrols and closed circuit television may need to be considered

2. Protection against falls

The Work at Height Regulations gives the following hierarchy of control:

- Avoid working at height, if possible
- Use an existing safe place of work
- Provide work equipment to prevent falls
- Mitigate distance and consequences of a fall
- Give instruction, training and supervision

Following a suitable risk assessment, the following hierarchy of measures should be considered:

- Avoid working at height, if possible
- Provide a properly constructed working platform, complete with toe boards and guard rails
- If this is not practicable or where the work is of short duration, suspension equipment should be used and only when this is impracticable
- Collective fall arrest equipment (air bags or safety nets) may be used
- Where this is not practicable, individual fall restrainers (safety harnesses) should be used
- Only when none of the above measures is practicable, should ladders or step ladders be considered

3. Fragile roofs

Particular hazards are:

- fragile roofing materials – more brittle with age and exposure to sunlight
- exposed edges
- unsafe access equipment
- falls from girders, ridges or purlins
- overhead services and obstructions
- the use of equipment such as gas cylinders and bitumen boilers
- manual handling hazards

Controls include:

- suitable means of access such as scaffolding, ladders and crawling boards
- suitable barriers, guard rails or covers where people work near to fragile materials and roof lights
- suitable warning signs indicating that a roof is fragile, at ground level

4. Protection from falling objects

- Use covered walkways or suitable netting to catch falling debris
- Waste material should be brought to ground level by the use of chutes or hoists
- Minimal quantities of building materials should be stored on working platforms
- Hard hats must be given to all employees whenever there is a risk of head injury from falling objects

Self-employed workers must supply their own head protection. Visitors to construction sites should be supplied with head protection and mandatory head protection signs displayed around the site

5. Demolition controls include

- prior to work, a full site investigation
- risk assessments (2 – project designer and contractor)
- a method statement
- waste disposal arrangements
- piecemeal and deliberate controlled collapse
- training and information to all workers

6. **Health hazards associated with construction activities**
   - noise, vibration, dust (including asbestos), cement, solvents, cleaners and biological agents
   - health surveillance
   - they must also be fenced and suitable notices posted

7. **Prevention of drowning**

   Arrangements must be in place to prevent people falling into the water and ensure that rescue equipment is available at all times.

8. **Vehicles and traffic routes**
   - All vehicles must be well maintained and only driven by trained persons
   - Traffic routes, loading and storage areas need to be well designed with enforced speed limits, good visibility and the separation of vehicles and pedestrians being considered
   - The use of one way systems and separate site access gates for vehicles and pedestrians may be required
   - The safety of members of the public must be considered particularly where vehicles cross public footpaths

9. **Fire and other emergencies**

   Emergency procedures required for fire, explosions, flooding or structural collapse should include:
   - the location of fire points and assembly points
   - accident reporting and investigation
   - rescue from excavations and confined spaces

10. Welfare facilities to include

- sanitary and washing facilities (including showers if necessary) with an adequate supply of drinking water
- accommodation for the changing and storage of clothes
- rest facilities for break times
- adequate first-aid provision (an accident book)
- protective clothing against adverse weather conditions

11. Electricity

- Only 110V equipment should be used on site
- If mains electricity is used, then residual current devices should be used with all equipment
- Where workers or tall vehicles are working near or under overhead power lines, either the power should be turned off or 'goal posts' or taped markers used to prevent contact with the lines
- Similarly, underground supply lines should be located and marked before digging takes place

12. Noise

- Noisy machinery should be fitted with silencers
- Ear defenders should be issued when working with noisy machinery
- When machinery is used in a workshop (such as woodworking machines), a noise survey should be undertaken

13. **Health hazards include**
    - dust – including asbestos
    - vibration
    - cement dust and wet cement
    - wood dust
    - solvents, cleaners and paints
    - tetanus and other biological agents
    - manual handling

14. **Waste disposal**
    - The disposal of waste must be properly managed
    - Waste skips must not be overloaded or used for inappropriate waste

The management of construction activities (CDM Regs)

The duty holders and their main duties, when the work is notifiable, are:

➤ The **Client** has a major role and must ensure that:

- all duty holders are competent
- pre-construction information is passed to relevant duty holders
- a CDM co-ordinator is appointed
- a Principal Contractor is appointed
- a health and safety plan is developed by the Principal Contractor
- a health and safety file is prepared
- adequate resources are available for the safe completion of the work

➤ The **Designer** must ensure that:

- adequate health and safety provision is incorporated into the design
- adequate information is provided about any significant risks associated with the design
- their work is co-ordinated with that of the other duty holders

➤ The **CDM co-ordinator** must:

- inform HSE on notifiable projects
- provide a link between duty holders, particularly with designers
- collect the pre-construction information and advise clients on any gaps
- manage the flow of information between clients, designers and contractors
- advise the client on the suitability of initial construction phase health and safety plan
- ensure that health and safety file is developed

➤ The *Principal Contractor* must:

- develop and manage the health and safety construction phase plan
- ensure that the construction phase is properly planned, managed and monitored
- employ only competent sub-contractors and supply them with necessary health and safety information
- ensure that suitable welfare facilities are provided on site
- display the project notification

Selection and control of contractors

See Element 3 of NGC1

Required documentation

1. Pre-construction information – needs to identify the hazards and risks associated with the design and construction work 2. Construction health and safety plan – defines the organization and arrangements required to control the site risks and co-ordinate the construction work	3. Health and safety file – is a record of health and safety information required by the subsequent users of the finished construction project 4. Notifiable work notice – issued to the HSE if the construction work is to last longer than 30 days or involves more than 500 person days of work

Working above ground level

Access equipment

Ladders

The following factors should be considered when using ladders:

- Aluminium ladders are light but should not be used in high winds or near live electricity
- Timber ladders need regular inspection for damage and should not be painted as this could hide cracks
- Ensure that the use of a ladder is the safest means of access for the work to be done and the height to be climbed
- The ladder needs to be stable in use with a safe inclination (1 in 4)
- The foot of the ladder should be tied to a rigid support
- The proximity of live electricity should be checked
- There should be at least 1 m of ladder above the stepping off point
- Over-reaching must be eliminated
- Workers who are to use ladders must be trained in the correct method of use and selection
- Ladders should be inspected (particularly for damaged or missing rungs) and maintained on a regular basis and they should only be repaired by competent persons

Certain work should not be attempted using ladders. This includes work where:

- two hands are required
- the work is at an excessive height
- the ladder cannot be secured or made stable
- the work is of long duration
- the work area is very large
- the equipment or materials to be used are heavy or bulky
- the weather conditions are adverse
- there is no protection from vehicles

Fixed scaffold

Fixed scaffolds are usually independently tied and important considerations are as follows:

- Scaffolding must only be erected and dismantled by competent people – any changes to the scaffold must be done by a competent person
- Adequate toe boards, guard rails and intermediate rails must be fitted to prevent people or materials from falling
- The scaffold must rest on a stable surface; uprights should have base plates and timber sole plates if necessary
- The scaffold must have safe access and egress
- Work platforms should be fully boarded with no tipping or tripping hazards
- The scaffold should be sited away from or protected from traffic routes so that it is not damaged by vehicles
- The scaffold should be properly braced and secured to the building or structure
- Overloading of the scaffold must be avoided
- The public must be protected at all stages of the work
- Regular inspections of the scaffold must be made and recorded

Components of a scaffold

- standard
- ledger
- guard rail
- toe boards
- bracing
- transom
- base plate
- sole board
- ties
- working platform

Mobile scaffold towers

The following points must be considered:

- The selection, erection and dismantling of mobile scaffold towers must be undertaken by competent persons
- Maximum height to base ratios should not be exceeded
- Diagonal bracing and stabilizers should be used
- Access ladders must be fitted to the narrowest side of the tower or inside the tower
- Persons should not climb up the frame of the tower
- All wheels must be locked whilst work is in progress
- All persons must vacate the tower before it is moved
- The tower working platform must be boarded, fitted with guard rails and toe boards and not overloaded

- Towers must be tied to a rigid structure if exposed to windy weather
- Persons working from a tower must not over-reach or use ladders from the work platform
- Safe distances must be maintained between the tower and overhead power lines both during working operations and when the tower is moved
- The tower should be inspected on a regular basis and a report made

Mobile elevated work platforms

The following factors must be considered when using mobile elevated work platforms:

- The mobile elevated work platform must only be operated by trained and competent persons
- It must never be moved in the elevated position
- It must be operated on level and stable ground
- The tyres must be properly inflated and the wheels immobilized
- Outriggers should be fully extended and locked in position
- Due care must be exercised with overhead power supplies and obstructions
- Procedures should be in place in the event of machine failure

Causes of accidents with access equipment

- ➤ design faults
- ➤ over-reaching
- ➤ falls and slips
- ➤ collision with vehicles
- ➤ unsound base support
- ➤ untrained personnel
- ➤ climbing with loads
- ➤ poor maintenance and/or inspection
- ➤ instability
- ➤ adverse wind effects

Inspections

- ➤ Every 7 days with a report
- ➤ Every day:
  - after any event likely to affect the stability of excavation
  - before the start of each shift
  - after an accidental fall of material

Excavations

Hazards:	Precautions:
➤ collapse of sides ➤ materials falling on workers ➤ falls of people and/or vehicles into excavation ➤ influx of groundwater ➤ underground services ➤ access and egress ➤ proximity of waste or stored materials or adjacent structures ➤ fumes and health hazards (Weil's disease)	➤ supervision by a competent person ➤ hard hats ➤ shore sides ➤ barrier around top ➤ well lit at night ➤ identify position of buried services ➤ care during the filling-in process